JUN'92 $ X 32.50

OXFORD STATISTICAL SCIENCE SERIES

SERIES EDITORS

J. B. COPAS   A. P. DAWID

G. K. EAGLESON   D. A. PIERCE

MARK J. SCHERVISH

S0-ABB-342

OXFORD STATISTICAL SCIENCE SERIES

Time Series
A Biostatistical Introduction

PETER J. DIGGLE

Department of Mathematics
Lancaster University

CLARENDON PRESS · OXFORD

Oxford University Press, Walton Street, Oxford OX2 6DP
Oxford New York Toronto
Delhi Bombay Calcutta Madras Karachi
Petaling Jaya Singapore Hong Kong Tokyo
Nairobi Dar es Salaam Cape Town
Melbourne Auckland
and associated companies in
Berlin Ibadan

Oxford is a trade mark of Oxford University Press

Published in the United States
by Oxford University Press, New York

© Peter J. Diggle, 1990

First published 1990
Reprinted 1991

All rights reserved. No part of this publication may be reproduced,
stored in a retrieval system, or transmitted, in any form or by any means,
electronic, mechanical, photocopying, recording, or otherwise, without
the prior permission of Oxford University Press

This book is sold subject to the condition that it shall not, by way
of trade or otherwise, be lent, re-sold, hired out or otherwise circulated
without the publisher's prior consent in any form of binding or cover
other than that in which it is published and without a similar condition
including this condition being imposed on the subsequent purchaser

British Library Cataloguing in Publication Data
Diggle, Peter J.
Time series: a biostatistical introduction.
1. Time series analysis
I. Title
519.5'5
ISBN 0–19–852226–6 (pbk)

Library of Congress Cataloging in Publication Data
Diggle, Peter.
Time series: a biostatistical introduction / Peter J. Diggle.
Includes bibliographical references.
1. Time-series analysis.   2. Biometry—Statistical methods.
I. Title
QA280.D54   1989   519.5'5—dc20   89-34018
ISBN 0–19–852226–6 (pbk)

Printed in Great Britain by
Biddles Ltd.,
Guildford and King's Lynn

To Linda

Preface

Time-series analysis is concerned with data which consist of time-ordered sequences of measurements on some phenomenon of interest. Most introductory books on time-series analysis take their principal motivation either from economics, with a strong emphasis on forecasting methods, or from signal processing in engineering and the physical sciences. My own motivation has been that of a statistician consulting in the biological sciences. One consequence of this is that the book's coverage of the subject, and more particularly the balance between topics, is slightly unconventional. Throughout the book, analyses of biological data sets are integrated into the methodological development. The data sets are listed in an appendix, to allow readers to reproduce my results for themselves, or to try alternative analyses.

Chapter 1 introduces the basic concepts of time-series analysis, including trend, serial dependence, and stationarity. Chapter 2 covers simple descriptive methods of analysis, including plotting, smoothing, estimation of the autocovariance function, and, finally, the periodogram and its connections with harmonic regression and with the estimated autocovariance function. Chapter 3 contains theoretical material on the properties of stationary random processes. Chapter 4 develops the ideas of spectral analysis, establishing and building on the link between the theoretical ideas in Chapter 3 and the more intuitively based data-analytic methods in Chapter 2. The discussion puts some emphasis on the analysis of short, replicated series which are common in biological applications, rather than on the long, unreplicated series which arise more often in engineering and the physical sciences.

Spectral analysis is essentially a non-parametric approach to the analysis of stationary time series. Chapter 5 describes an alternative, parametric approach which is particularly appropriate when the primary objective is an analysis of trends. I again emphasize methods for the analysis of short, replicated series, possibly involving several different experimental treatments. Data of this kind are often referred to in the biometrical literature as 'repeated measurements'. Chapter 6 considers the parametric approach for a class of models known as autoregressive integrated moving average processes, or ARIMA processes. In Chapter 7, results for ARIMA processes are incorporated into a discussion of forecasting. Chapter 8 is a brief introduction to the analysis of bivariate time series.

Suggestions for further reading are given at the ends of several chapters. These are intended to help readers to explore the literature in more detail. They do not constitute a formal bibliography.

I have assumed that readers have a knowledge of basic probability theory, statistical methods, matrix algebra, and calculus. Some previous exposure to the theory of stochastic processes, and to the multivariate Normal distribution, would be helpful but is not essential. Two further requirements, one statistical the other mathematical, are the matrix approach to the general linear model and the elementary manipulation of complex numbers. Since these topics are more specialized than the other prerequisites, I have given outlines of the required results in two appendices.

Most of the work on the book was done while I was employed in the Division of Mathematics and Statistics, Commonwealth Scientific and Industrial Research Organisation, Australia. I am indebted to many colleagues there for stimulating discussions on all manner of applied statistical research problems. I am especially grateful to Murray Cameron, Geoff Eagleson, and Nick Fisher for their valuable comments on early drafts of the book. I also thank the contributors of data sets, who are acknowledged individually in Appendix A. Last but not least, Coleen Walters, Kaye Nankivell, and Helen Kaminsky performed miracles with an almost illegible manuscript.

Lancaster   P.J.D.
November 1988

Contents

1
Introduction

1.1 Definitions, examples, and notation

In its simplest form, a *time-series* is no more than a set of data $\{y_t : t = 1, \ldots, n\}$ in which the subscript t indicates the *time* at which the datum y_t was observed. A number of elaborations on this basic theme are of course possible, including the following.

- The points in time at which the observations are taken need not be equally spaced. A more honest notation for the data would then be $\{y(t_i) : i = 1, \ldots, n\}$.
- Each datum may represent an accumulation of some underlying quantity over an interval of time, rather than its value at a single point, for example daily rainfall.
- The data set may be augmented by replicate series. For example, we might monitor the body weight of each of a number of experimental animals over time.
- Each scalar quantity y_t might be replaced by a vector $\mathbf{y}_t = (y_{1t}, \ldots, y_{pt})$ giving the values of p quantities which are in some way related. For example, each $\mathbf{y}_t$ might represent a daily reading of the temperature, blood pressure, and pulse rate of a hospital patient.

Most statistical methods encountered in elementary textbooks assume that the individual observations which make up a set of data are realizations of *mutually independent* random variables. Typically, the assumption of mutual independence is justified by careful attention to various aspects of the experimental procedure, including the random selection of a sample of experimental units from some larger population, the random allocation of experimental units to experimental treatments, and, most pertinently from our point of view, the taking of a single observation from each experimental unit.

When a sequence of observations is taken on each experimental unit, the assumption of mutual independence is seldom sustainable. The particular flavour of time-series analysis is its concern with the nature of the *dependence* amongst the members of a *sequence* of random variables. Some examples serve to illustrate these very basic ideas, and to identify some of the objectives of time-series analysis.

Example 1.1. An electrocardiogram trace

Figure 1.1 shows an essentially continuous ECG trace, $y(t)$, say, taken from a healthy adult female. The overall impression is of a very regular, repeating pattern. Each cycle, corresponding to the time span between successive heartbeats, exhibits a number of separately identifiable features. The small rise and fall, labelled P on the inset to Fig. 1.1, is due to electrical activity in the atrium. The following sharp rise and fall, labelled QRS, represents electrical activity in the ventricle. The subsequent asymmetric rise and fall, labelled T, reflects repolarization of the ventricle. Between each PQRST episode there is a quiescent period in which $y(t)$ is approximately constant. Closer inspection reveals apparently random variation in the basic pattern. For example, the heights of the peaks and the separations in time between successive peaks vary between cycles. Also, the basal level of the quiescent period shows a noticeable fall during the early part of the trace, with an equally noticeable increase later on.

In order to analyse these data, we have to sample them at discrete time points. The choice of sampling interval is clearly important: too-frequent sampling is potentially wasteful in terms of both storage of information and subsequent computation, whereas too-infrequent sampling risks losing some of the essential features of the original trace. For example, Fig. 1.2 shows the series $\{y_t : t = 1, \ldots, 80\}$ obtained by sampling the initial portion of the trace in Fig. 1.1 at unit time intervals. The basic cyclic pattern is still just discernible, but its magnitude is greatly diminished because the discrete-time sampling misses most of the sharp peaks of the QRS phases. Also, the finer structure of each PQRST episode is lost entirely.

Example 1.2. Levels of lutenizing hormone (LH) in blood samples

In this example, blood samples have been taken from a healthy woman at ten-minute intervals over an eight-hour period. The samples are later assayed to determine the corresponding LH levels, which then constitute the data for analysis.

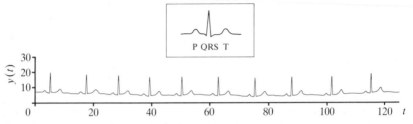

Fig. 1.1. An ECG trace from a healthy adult female. The inset shows an enlargement of a single heartbeat.

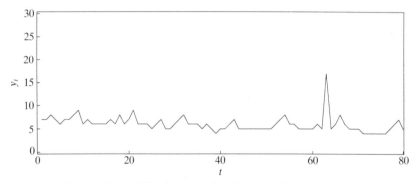

Fig. 1.2. A time series of 80 values obtained by sampling the intitial portion of the ECG trace in Fig. 1.1 at unit time intervals.

Figure 1.3 shows three series, one from the early follicular phase of the subject's menstrual cycle, the others from the late follicular phase of two successive cycles. Note that one of the series contains a gross outlier, presumably the result of a recording error or a failure of the assay. In all subsequent analyses of these data, we replace the outlier by the interpolated value shown by a dashed line in Fig. 1.3c. As with the ECG data, the phenomenon being investigated exists at all times, but in this case continuous monitoring is impractical. The ten-minute interval was chosen after an earlier analysis of series taken at five-minute intervals had suggested that the important features of the pattern of variation in LH levels would be retained using the longer sampling interval – the necessity for limiting the total amount of blood removed is obvious!

LH plays an important role in the reproductive system. The fundamental questions posed by these data include the investigation of possible cyclic patterns within each series due to pulsatile release of LH, characterization of differences between the properties of series taken from different phases of the menstrual cycle, and assessment of whether or not series taken from the same phase of successive cycles are comparable. One point of entry to the medical literature on pulsatile release of LH is Murdoch *et al.* (1985).

Example 1.3. Wool prices at weekly markets
This example relates to wool prices as monitored by the Australian Wool Corporation over the period July 1976 to June 1984 inclusive. Prices are monitored weekly, except for certain weeks in which markets are not held; for example, there is a break of several weeks each Christmas. Immediately prior to each week's markets, the Corporation set a floor price. This determined the Corporation's policy on intervention buying and is therefore a reflection of the overall price of wool for the week in

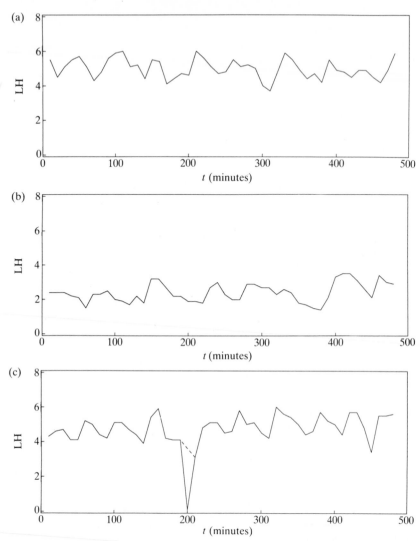

Fig. 1.3. Three time series of concentrations of luteinizing hormone in blood samples taken from a healthy woman at ten-minute intervals over an eight-hour period. (a) Late follicular phase of first menstrual cycle. (b) Early follicular phase of second menstrual cycle. (c) Late follicular phase of second menstrual cycle.

question. The actual prices paid for different grades of wool nevertheless show considerable variation about the floor price. Figure 1.4 shows two series, the floor price and the actual price paid for fine grade wool (19 μm nominal thickness) over the nine years in question. Comparable series are available on the prices paid for other grades.

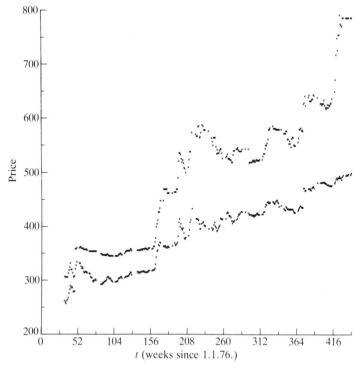

Fig. 1.4. Two time series of Australian wool prices at weekly markets during the period July 1976 to June 1984. The upper series represents the price paid for fine grade wool (19 μm nominal thickness), and the lower series the floor price set by the Australian Wool Corporation.

Both series in Fig. 1.4 show clear seasonal fluctuations about a generally rising trend which includes one or two fairly abrupt changes. One such abrupt change spans the devaluation of the Australian dollar in March 1983. This suggests that the series need to be adjusted to take account of currency fluctuations. One simple form of adjustment is to analyse the ratio of actual price paid to floor price. This would also remove long-term trends due to inflation and gross seasonal effects, which may well be desirable if, for example, the primary interest is in the relative price movements for different grades of wool.

A further consideration is that price movements are typically multiplicative in nature, suggesting a logarithmic transformation prior to analysis of the data. We might therefore choose to analyse series of derived quantities,

$$y_t = \log(\text{price paid/floor price}).$$

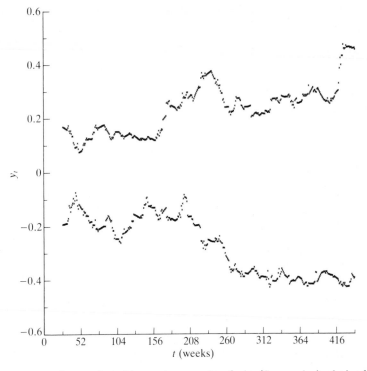

Fig. 1.5. Two time series of log ratios, $y_t = \log$ (price/floor price), derived from the wool price data. The upper series corresponds to fine grade wool (19 μm nominal thickness) and the lower series to coarse grade wool (30 μm nominal thickness).

Figure 1.5 shows two such series, for fine (19 μm) and coarse (30 μm) grades of wool, and highlights the very different trends in the relative prices paid for the two grades of wool over the nine-year period.

A comparison between Examples 1.2 and 1.3 allows us to draw a distinction between the adjectives seasonal and cyclic as applied to time-series data. By the former, we mean any periodic behaviour which is known *a priori*: thus, for example, we would take it to include annual variation in daily air temperature readings and daily variation in hourly air temperature readings. By cyclic, we mean any periodic behaviour which may or may not be known in advance: for example, it is well known that LH levels in blood samples show some degree of cyclic variation within a time-span of several hours, but the precise cycle length is not obvious *a priori*, and indeed appears to vary both between and, on a sufficiently long time-scale, within individuals.

To put it briefly: cyclic embraces seasonal, but not *vice versa*.

Example 1.4. UK deaths from bronchitis, emphysema, and asthma
Figure 1.6 shows the monthly returns of deaths in the United Kingdom
attributed to bronchitis, emphysema, and asthma over the years 1974–9
inclusive: returns for males and for females are shown separately. The
dominant features of the data are the large discrepancy between the
overall returns for males and for females, the strong seasonal pattern and
the unusually high returns for the winter of 1975–6 (months 24–6). Note
also that, in contrast to our earlier examples, each observation represents
an accumulation over an interval of time, here one month, rather than an
instantaneous value.

At one level, there is an obvious relationship between the male and
female returns, namely their common seasonal pattern. It would be of
some interest to determine the extent to which the two series of
deviations from the seasonal pattern are also related. A strong relationship at this level might imply that the residual fluctuations in monthly
returns, after correcting for seasonal variation, are largely attributable to
a common environmental factor – plausibly, variations in weather patterns about the prevailing climatic trend, which is undoubtedly the source
of the seasonal variation in the data. Conversely, a weak relationship
would be more suggestive of random variation in the inherent suscep-
tibility of the individuals at risk in a particular month.

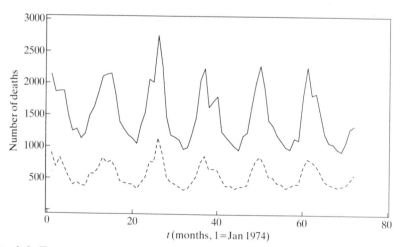

Fig. 1.6. Two time series of monthly returns of deaths in the United Kingdom
attributed to bronchitis, emphysema, and asthma over the years 1974 to 1979.
—, Males; – – –, females.

Example 1.5. Growth of colonies of the protozoon, *paramecium aurelium*

Gause (1934) describes a series of experiments concerning the growth of closed colonies of *paramecium aurelium* in a nutritive medium which consists of a suspension of the bacterium *bacillus pyocyaneous* in salt solution. At the beginning of each experiment, 20 individuals were placed in a tube containing 5 ml of the medium at a constant temperature of 26°C. After two days, the tube was carefully stirred, a sample of 0.5 ml taken and the number of individuals counted. After centrifuging of the remaining 4.5 ml, the medium was drawn off and the residue washed with bacteria-free salt solution to remove accumulated waste products. After a second centrifuging and drawing off of the liquid, fresh medium was added to restore the volume to 5 ml. The whole procedure was repeated daily until day 25, and the entire experiment replicated three times.

Figure 1.7 shows the three sets of counts from 0.5 ml samples, together with the means of the three replicates. A natural question to ask of these data is whether we can develop a simple model for the population mean count as a function of time. In fitting any such model we should nevertheless recognize that successive sample counts within each replicate are likely to be statistically dependent.

Example 1.6. Body weights of rats under three different experimental treatments

Box (1950) gives the results of an experiment in which 30 rats were assigned at random to three treatment groups. Treatment 1 was a

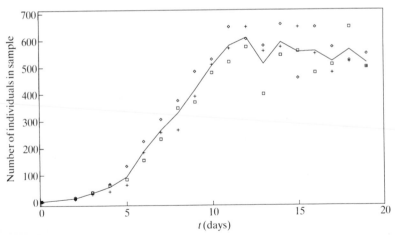

Fig. 1.7. Data on the growth of colonies of *paramecium aurelium*. +, First replicate; □, second replicate; ◇, third replicate; —, mean of three replicates.

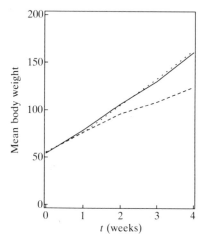

Fig. 1.8. Mean response profiles for body weights of rats in an experiment to determine the effects of two different additives to the rats' drinking water. —, Control (no additive); - - - -, thyroxin; –––, thiouracil.

control. Treatments 2 and 3 consisted of two different additives to the rats' drinking water (thyroxin and thiouracil respectively). The data for each rat consist of the body weight at the beginning of the experiment ($t = 0$) and at times $t = 1$, 2, 3, and 4 weeks later; results for three of the rats on treatment 2 are missing due to unspecified accidents. Figure 1.8 shows the mean response profile within each of the three treatment groups.

As in Example 1.5, the questions of primary interest for these data concern the mean response profiles for some underlying population of rats and, more particularly, the effects of the experimental treatments 2 and 3 on these profiles. The structure of the random variation about the mean response is of interest only insofar as it affects the appropriateness of any particular method for making inferences about the mean response profiles in the different treatment groups.

With as few as five observations on each experimental unit, one might question whether there is anything to be gained by explicit recognition of the time-series aspect of these data. Why not simply regard each response profile as an arbitrary, five-dimensional random variable? The justification for at least contemplating a time-series approach lies in the resulting economy of expression: the variance structure of an arbitrary five-dimensional random variable involves 15 parameters (five variances and 10 covariances between distinct pairs), whereas a typical time-series model might involve three or four parameters at most. It is a quite

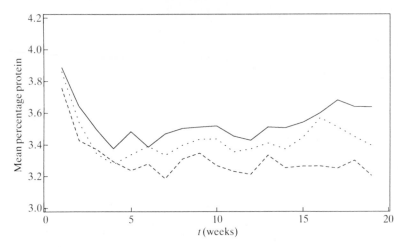

Fig. 1.9. Mean response profiles for protein content of milk samples from cows in an experiment comparing the effects of three different diets. —, Barley; - - - -, barley and lupins; –––, lupins.

general statistical principle that economical parameterizations, if justified, enable more efficient inferences (Altham 1984). Clearly, this argument becomes more compelling when applied to longer series.

Example 1.7. Protein content of milk samples
The data for this example relate to the protein content of milk samples from each of 79 cows. Samples were taken weekly for 19 weeks. However, about half of these series of measurements were terminated prematurely because the cows in question stopped producing milk, and a few series have isolated missing values at earlier times. The cows were allocated to one of three diets: barley, lupins, or a mixture of the two. Figure 1.9 shows the observed mean response profile over time in each treatment group.

As with Example 1.6, the main focus of attention for these data is the question of whether the mean response profiles differ amongst the three experimental treatments. However, because the series of measurements are longer (up to 19, rather than 5) and many of the series are incomplete, a time series approach to the analysis of the data is more likely to pay dividends.

1.2 Objectives of time-series analysis

A modest objective of any time-series analysis is to provide a concise *description* of a historic series. The description may consist of a few summary statistics, but is more likely to include one or more graphical representations of the data. This is because the complexity of a time-series, as opposed to a random sample from a distribution, is often such as to require a *function*, rather than a single number, to summarize each of its essential features; a simple example is the description of the mean value by a function of time, $\mu(t)$ say, rather than by a single quantity μ.

A more ambitious task is to *forecast* future values of a series. Indeed, much time-series analysis has been developed to this specific end, with its obvious relevance to sales forecasting and other economically oriented applications; see, for example, Box and Jenkins (1970, especially Chapter 5), or Gilchrist (1976).

A problem which arises naturally in medicine and elsewhere is the *monitoring* of a time series to detect changes in behaviour as they occur. A good example is given by Smith and West (1983), who develop a methodology for on-line detection of deteriorating renal function in post-operative renal transplant patients.

A more passive, but nevertheless important, role for time-series analysis is to *accommodate* serial dependence in the course of making inferences about basic structural parameters. For example, the data in Example 1.7 were collected with the primary purpose of establishing and describing the nature of the differences between the experimental treatments. The time-series aspect of the data is incidental to this question, but has a bearing on the development of valid and efficient inferential procedures.

Finally, we distinguish two different types of modelling for time-series data. Models may be developed either with a view to furthering a scientific understanding of the underlying mechanisms which generate the data, or with the more pragmatic aim of helping towards the attainment of previously stated objectives.

1.3 More notation

Throughout this book, random variables are denoted by upper case letters and their realized values by the corresponding lower case letters. A lower case t always denotes time. Thus, a *time series* $\{y_t : t = 1, \ldots, n\}$ is to be understood as a set of realized values of random variables $\{Y_t : t = 1, \ldots, n\}$. Sometimes, for example when data are collected at irregularly spaced time points, we shall want to emphasize the continuous nature of time. We shall then write $Y(t)$ rather than Y_t and similarly, $y(t)$ rather than y_t.

All observed time series are necessarily finite in extent, but for their theoretical description it is usually convenient to regard them as infinitely extendible. This leads us to study the properties of infinite *random sequences*, which we write simply as $\{Y_t\}$, and *random functions* $\{Y(t)\}$. The distinction between these is that in the former case, t assumes only integer values, whereas in the latter it varies over the whole real line. We shall use the term *random process* to mean either a random sequence or a random function, according to context.

The distinction between discrete and continuous time has theoretical implications which, in practice, are often ignored. Here, and for future reference, we note three different ways in which a random sequence $\{Y_t\}$ can arise.

(a) The underlying time-scale may be genuinely discrete, in that the phenomenon of interest does not exist at intermediate times (cf. Example 1.3, at least with regard to the floor prices which are set in advance of each week's markets).

(b) An underlying random function $\{X(t)\}$ may be sampled at equally spaced time points (cf. Example 1.2).

(c) An underlying random function $\{X(t)\}$ may be accumulated over equal intervals of time, so that

$$Y_t = \int_{t-1}^{t} X(s)\,\mathrm{d}s,$$

(cf. Example 1.4).

A second distinction is between *discrete-valued* and *continuous-valued* random variables at each time point. Most of the methods described in this book are appropriate to time-series data in which each observation is measured on an essentially continuous scale. All of our examples from 1.1 to 1.7 are of this type. Examples 1.4 and 1.5 strictly concern discrete-valued observations, but the numbers involved are sufficiently large that little or nothing is lost by treating them as continuous-valued.

Much of the methodology for discrete-valued time-series data has been developed to meet the requirements of relatively specialized problems. One example is the modelling of daily rainfall data. Gabriel and Neumann (1962) use a two-state Markov chain to describe the sequence of wet and dry days at a particular location. In this model, the probability that tomorrow will be dry, given the preceding sequence of wet and dry days, is assumed to depend only on whether today is wet or dry. Stern and Coe (1984) describe a more general model, which allows for higher-order dependence in the sequence of wet and dry days, a continuous component for the amount of rain on wet days and seasonal variation in the parameters which govern the behaviour of the model.

A rather different type of discrete-valued time series is a series of events. Here, the data comprise a sequence of stochastically generated times at which events of particular interest occur. For example, the events in question might be the times of cell division in a process of cellular proliferation. One way to record such data is in the form of counts in successive equal time intervals, defining an accumulated random sequence (cf. Example 1.4); another is as a sequence of the time intervals between successive events, defining a continuous-valued random sequence $\{Y_t\}$ in which time is genuinely discrete, albeit not conventional clock time but merely an identifier for the ordering of the sequence. Statistical methods for the analysis of series of events are described in Cox and Lewis (1966).

1.4 Trend, serial dependence, and stationarity

The concepts of trend, serial dependence and stationarity are central to any understanding of the probabilistic structure of time-series data. For definiteness, we shall describe these three concepts in the context of random functions $\{Y(t)\}$, although the ideas apply also to random sequences.

Firstly, the non-random function $\mu(t) = E[Y(t)]$, where $E[\cdot]$ denotes expectation or mean value, is called the *trend* of $\{Y(t)\}$.

Secondly, the term *serial dependence* refers to the fact that the random variables $Y(t)$ and $Y(s)$ are statistically dependent for at least some pairs of values (s, t) with $s \neq t$. In particular, $Y(t)$ and $Y(s)$ are typically correlated, and this leads us to define the *autocovariance function* of $\{Y(t)\}$ as

$$\gamma(t, s) = E[\{Y(t) - \mu(t)\}\{Y(s) - \mu(s)\}].$$

Thirdly, the idea of stationarity is that the probabilistic structure of $\{Y(t)\}$ is unaffected by a shift in the time origin; put rather more loosely, $\{Y(t)\}$ looks the same at whatever point in time we begin to observe it. A formal definition of strict stationarity is that for any positive integer m and times $t_i : i = 1, \ldots, m$, the joint probability distribution of $\{Y(t_i + s) : i = 1, \ldots, m\}$ is the same for all values of s.

Strict stationarity is a strong and, in practice, uncheckable assumption. A weaker assumption is that of second-order stationarity, whereby $\mu(t)$ is equal to a constant, μ, for all t and $\gamma(t, s)$ depends only on $|t - s|$, the absolute difference between t and s. In this book, the word stationary used without qualification means second-order stationary.

The simplest possible example of a stationary random sequence is *white noise*. This consists of a sequence of mutually independent random variables, each with mean zero and finite variance σ^2. Its autocovariance function is

$$\gamma(t, s) = \begin{cases} \sigma^2 & : & t = s \\ 0 & : & t \neq s. \end{cases}$$

We shall use white noise as a building block from which to construct more interesting random sequences. From a data-analytic viewpoint, white noise provides a benchmark against which we can assess possible serial dependence in an observed time series. Figure 1.10 shows a simulated realization of each of two white noise processes. In Fig. 1.10a, the observations are Normally distributed with probability density

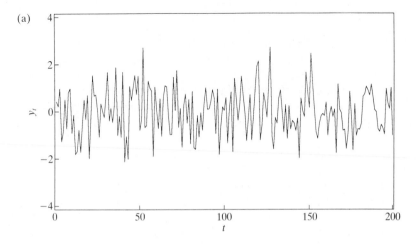

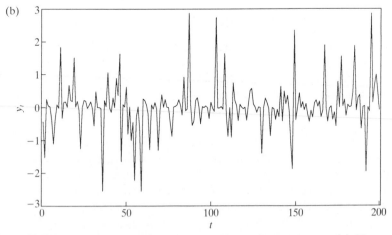

Fig. 1.10. Simulated realizations of two white noise processes. (a) Normally distributed. (b) Gamma distributed.

proportional to $\exp(-y^2/2)$, whereas in Fig. 1.10b they follow a signed gamma distribution with probability density proportional to $|y|^{-0.5}\exp(-|y|)$. The different shapes of the two densities are reflected in the appearances of the two simulated series; in particular, the sharp spikes in Fig. 1.10b reflect the fact that the signed gamma density has a relatively high concentration of probability close to zero but also relatively long tails. Incidentally, the reason for using the term white noise will become clear in Chapter 3.

A pragmatic view of stationarity is that it provides a means of manufacturing some degree of replication within a single time series, thereby making formal inference possible. A theoretical framework within which we can handle a very wide range of practical problems is a model of the form

$$Y(t) = \mu(t) + U(t), \tag{1.4.1}$$

where each $Y(t)$ represents the measurement to be made at time t, or some transformation thereof, $\mu(t)$ is the trend of $\{Y(t)\}$ and $\{U(t)\}$ is a stationary random function.

The following is an example of the type of transformation which might be required to bring a time series within the ambit of eqn. (1.4.1).

Example 1.8. A simple random walk
Let Y_0 be identically zero and define a random sequence $\{Y_t\}$ by the recursion

$$Y_t = Y_{t-1} + Z_t \quad : \quad t = 1, 2, \ldots,$$

where $\{Z_t\}$ is white noise.

The sequence $\{Y_t\}$ so defined is an example of the type of stochastic process known as a *random walk*. Let $\mu_t = E[Y_t]$. Using elementary properties of expectation, we can write

$$\mu_t = \mu_{t-1} + E[Z_t] = \mu_{t-1}.$$

But clearly, $\mu_0 = 0$ and we conclude that $\mu_t = 0$ for all t. This satisfies one of the requirements for $\{Y_t\}$ to be stationary. However, a similarly elementary calculation shows that the variance of Y_t is $\gamma(t, t) = t\sigma^2$, which violates the requirement for stationarity that $\gamma(t, s)$ should depend only on $|t - s|$. We have therefore demonstrated that $\{Y_t\}$ is a *non-stationary* random sequence. However, the sequence of differences between successive Y_t identifies the white noise sequence $\{Z_t\}$ which *is* stationary.

Figure 1.11 shows a simulated realization of a simple random walk, of length 200 observations. Note how the phenomenon of variability increasing with time generates a false impression of trend because the realized values can drift progressively further from zero.

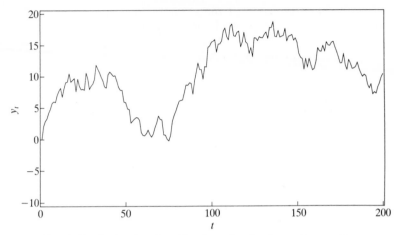

Fig. 1.11. A simulated realization of a simple random walk.

1.5   Duality between trend and serial dependence

Equation (1.4.1) expresses a decomposition of $Y(t)$ into a non-random function $\mu(t)$ and a random function $U(t)$. This is all very well in theory, but the following logical difficulty arises in practice. Any realization of $U(t)$ is itself a non-random function $u(t)$. It follows that in a single time series $\{y(t_i):i = 1, \ldots, n\}$, however long, we cannot hope to separate $\mu(t)$ and $u(t)$ unless we are prepared to make further assumptions. In practice, most analyses of time-series data attempt to do precisely this, either by the explicit declaration of a parametric model for $\mu(t)$ or $\{U(t)\}$, or by an implicit assumption that $\mu(t)$ is 'smooth' whilst $u(t)$ is 'rough'. Judgement on what is smooth and what is rough may, however, depend on the scale of observation. An example makes the point.

Example 1.9. A smooth random function
Figure 1.12 shows a partial realization of a stationary random function $\{Y(t)\}$ with

$$\gamma(t, s) = (4\pi)^{-1/2} \exp\{-0.25(t - s)^2\}.$$

Its precise construction is that

$$Y(t) = \sum_{i=1}^{\infty} f(t - T_i),$$

where $f(t) = (2\pi)^{-1/2} \exp(-t^2/2)$ is the standard Normal density function and the T_i form a homogenous Poisson point process with unit rate, i.e.

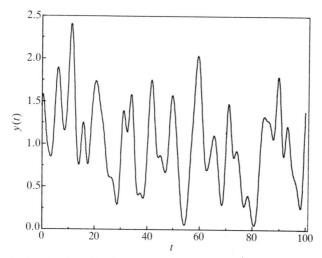

Fig. 1.12. A simulated realization of a stationary random function on the interval $0 \leq t \leq 10$.

the time intervals between successive T_i are an independent random sample from an exponential distribution with mean 1.0 (Cox and Lewis, 1966, Chapter 2). The erratic fluctuations in the realization $y(t)$ do indeed suggest a random function. Figure 1.13, on the other hand, looks far from random, yet it is simply a plot of the same $y(t)$ but restricted to the range $0 \leq t \leq 10$.

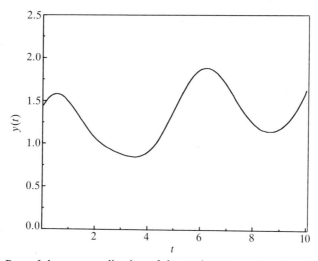

Fig. 1.13. Part of the *same* realization of the stationary random function shown in Fig. 1.12, but restricted to the interval $0 \leq t \leq 10$.

The practical message of Example 1.9 is that a data-based decision to treat particular aspects of a single time series as random or non-random has more to do with convenience than with absolute truth. Parametric models which typically combine random and non-random aspects must similarly be viewed as conveniences unless they are based on scientific reasoning external to the data in question. None of this denies the usefulness of models whose validity cannot be established formally. Purists may, however, decide at this point that time-series analysis is not for them!

1.6 Software

Good software for time-series analysis is becoming increasingly widely available. Sources include general-purpose statistical packages such as GENSTAT (Payne 1987) or SPSS-X (SPSS 1988), subroutine libraries such as IMSL or NAG and a variety of more specialized software products. Most of the analyses in this book were carried out using the S system (Becker *et al.* 1988), which provides an excellent environment for interactive data analysis and statistical graphics. Another feature of S which makes it attractive to the professional statistician is the ease with which the system can be extended to meet the requirements of particular groups of users.

2
Simple descriptive methods of analysis

2.1 Time plots

The most basic graphical representation of a time series $\{y(t_i):i=1,\ldots,n\}$ is as a *time plot* of the observed values $y(t_i)$ against the times of observation t_i. Often, the display will be enhanced if we join successive points on the graph by straight lines. This is particularly the case if we want to compare several time series on the same graph, since a picture consisting of traces formed by different line types is usually easier to digest than one consisting of scatter plots which use different plotting symbols. For example, Fig. 2.1 contains precisely the same information as Fig. 1.6, but conveys its message less effectively.

A counter-argument is that joining points gives a false impression of continuous observation. The zigzag appearance of the trace made by the joined points is an effective visual cue against this wrong interpretation, but the objection has some validity, particularly for series with missing

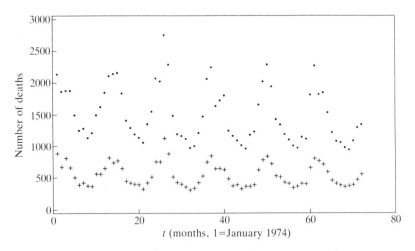

Fig. 2.1. Two time series of monthly returns of deaths in the United Kingdom attributed to bronchitis, emphysema, and asthma over the years 1974 to 1979. Male deaths are shown as dots, female deaths are crosses.

values. For example, by not joining successive points in Figs 1.4 and 1.5 we drew the reader's attention to the gaps and jumps in the series which are an intrinsic feature of the data.

Another choice which can affect the interpretation of a time plot is its aspect ratio, or 'shape parameter', as discussed by Cleveland *et al.* (1988). Figure 2.2 shows two time plots of a well-known set of data on the annual numbers of lynx trapped in the Mackenzie River district of north-west Canada from 1821 to 1934. The data show a strong, approximately ten-year cycle. In Fig. 2.2a we see clearly the differences between the heights of successive peaks. In Fig. 2.2b we have compressed the *y*-axis relative to the *x*-axis. The effect of this is to highlight the asymmetry of the rise and fall within each cycle: most cycles show a relatively slow increase followed by a faster decline. A number of different statistical descriptions of the ten-year cycle in these data have appeared over the years. See, for example, Moran (1953), Bulmer (1974), Campbell and Walker (1977), and Tong (1977).

Tufte (1983) contains a wealth of good advice and insight concerning these and other aspects of the graphical presentation of data.

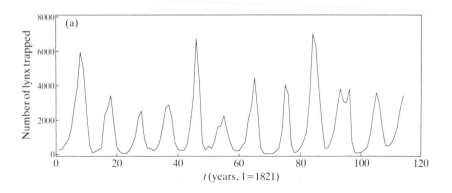

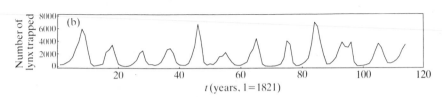

Fig. 2.2. Annual numbers of lynx trapped in the Mackenzie River district of north-west Canada from 1821 to 1934. (a) Aspect ratio chosen to emphasize differences between the heights of successive peaks. (b) Aspect ratio chosen to emphasize the asymmetry of the rise and fall within each cycle.

2.2 Smoothing

Figure 2.3a shows a time plot of the 'female' data from Example 1.4. The raw data are shown as unconnected symbols. The superimposed continuous trace is formed not by joining successive data points (t, y_t), but rather by joining points (t, s_t), where each s_t is defined as

$$s_t = (y_{t-1} + y_t + y_{t+1})/3 \qquad (2.2.1)$$

(a)

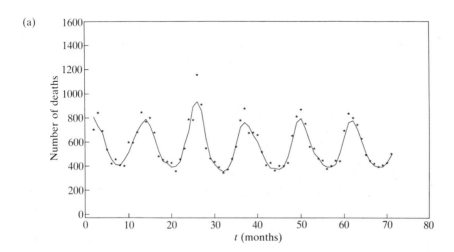

(b)

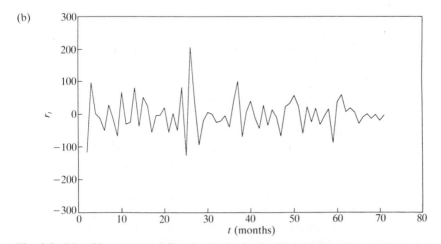

Fig. 2.3. Monthly returns of female deaths in the United Kingdom attributed to bronchitis, emphysema, and asthma over the years 1974 to 1979. (a) Raw data, y_t, shown as asterisks, and smoothed data, $s_t = (y_{t-1} + y_t + y_{t+1})/3$, shown as connected line segments. (b) Residuals, $r_t = y_t - s_t$.

Figure 2.3a emphasizes the major features of the data, notably the seasonal pattern, whilst de-emphasizing the apparently random fluctuations which feature in the time plot of the raw data. On the other hand, Fig. 2.3b, which is a time plot of the 'residuals', $y_t - s_t$, re-emphasizes the random aspects of the data without the distraction of the seasonal pattern.

More generally, we use the term *smoothing* to mean a decomposition of a time series y_t into a 'smooth' component s_t and a 'rough' component r_t, so that

$$y_t = s_t + r_t. \tag{2.2.2}$$

As it stands, eqn. (2.2.2) is vacuous. In order to give it some substance, we might be tempted to relate it to the basic model expressed as eqn. (1.4.1),

$$Y(t) = \mu(t) + U(t),$$

suggesting that s_t is an estimate of the trend, $\mu(t)$. This formulation is useful for motivating a number of specific methods of smoothing, but should not be taken too literally. Smoothing is fundamentally an exploratory operation, a means of gaining insight into data *without* precisely formulated models or hypotheses. For the time being, we take it that the terms 'smooth' and 'rough' have an obvious intuitive meaning, and press on.

2.2.1 *Moving averages*

The smoothing operation defined by eqn. (2.2.1) consists of averaging three successive observations centred on the time point of current interest. In general, a *moving average* of a time series $\{y_t : t = 1, \ldots, n\}$ is a time series $\{s_t\}$ defined by

$$s_t = \sum_{j=-p}^{p} w_j y_{t+j} \quad : \quad t = p+1, \ldots, n-p, \tag{2.2.3}$$

where p is a positive integer and the w_j are *weights*, with $\sum w_j = 1$. Typically, each w_j is positive and $w_j = w_{-j}$. We call $2p+1$ the *order* of the moving average. Note that eqn. (2.2.3) leaves s_t undefined near the ends of the series. One way to extend the definition is to let the summation in eqn. (2.2.3) range from $j = \max(-p, 1-t)$ to $j = \min(p, n-t)$ and divide by the corresponding sum of the included weights. In practice, p is usually very much smaller than n and little is then lost by leaving the observations near the ends of the series unsmoothed. The restriction to moving averages of odd order is slightly artificial, but is imposed in order to preserve an unambiguous correspondence between y_t and s_t in eqn. (2.2.2).

How should the order and the weights be chosen? We have already given one example of an equally weighted, or *simple* moving average of order 3, namely eqn. (2.2.1). One way to build up a system of unequally weighted moving averages is to use two or more successive applications of a simple moving average of order 3. For example, two successive applications would give

$$s_t = \{(y_{t-2} + y_{t-1} + y_t)/3 + (y_{t-1} + y_t + y_{t+1})/3 + (y_t + y_{t+1} + y_{t+2})/3\}/3$$
$$= (y_{t-2} + 2y_{t-1} + 3y_t + 2y_{t+1} + y_{t+2})/9.$$

This approach gives weights with the intuitively attractive property that they fall off symmetrically about a maximum, i.e. $w_j = w_{-j}$ and $w_0 > w_1 > \cdots > w_p$. It also smooths to some extent all the observations except y_1 and y_n, although not in the manner previously suggested. Furthermore, the method of construction makes it clear that each successive application imparts a greater degree of smoothness to the series $\{s_t\}$.

Sometimes, the particular context of the data can suggest a suitable moving average. For example, the data from Example 1.4 would be expected *a priori* to show seasonal fluctuations. If we want a moving average which is free of this seasonal effect, for example to highlight possible long-term trends, we should choose a set of weights such that each of the twelve months of the year makes an equal contribution. The simplest example would be a moving average of order 13, with $w_6 = w_{-6} = \frac{1}{24}$ and all other $w_j = \frac{1}{12}$. Conversely, if we want a moving average which highlights the seasonal effect, a much lower order would be appropriate, for example the simple moving average of order 3 used to produce Fig. 2.3a.

Example 2.1 UK deaths from bronchitis, emphysema, and asthma
Figure 2.4 shows the result of applying each of the above two moving averages to the male and female data of Example 1.4. An obvious feature of the three-point moving average smoothing, also evident in the raw data plot (Fig. 1.6), is the unusually large number of deaths in the winter of 1975–6. The thirteen-point moving average reveals a gentle downward trend in the numbers of male deaths over the six year period covered by the data.

2.2.2 *Polynomial regression*

An alternative approach to smoothing is to fit a polynomial to the data (t_i, y_i), where $y_i = y(t_i)$. This treats smoothing as a regression problem in which y is the response and integer powers of t are the explanatory

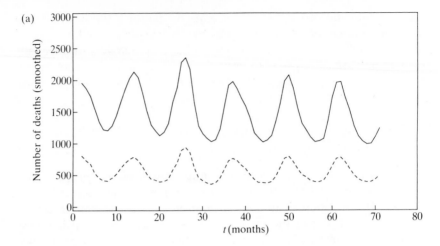

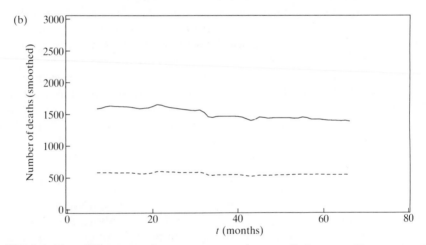

Fig. 2.4. Two different moving average smoothers applied to monthly returns of deaths in the United Kingdom attributed to bronchitis, emphysema, and asthma. ——, males; – – –, females. (a) Three-point moving average. (b) Thirteen-point moving average.

variables. The resulting smooth function is of the form

$$s(t) = \sum_{j=0}^{p} b_j t^j, \qquad (2.2.4)$$

where the b_j are to be estimated from the data.

The usual method of estimating the b_j is to minimize the quantity $\sum_{i=1}^{n} \{y_i - s(t_i)\}^2$. Using results from Appendix B, we can write the solution explicitly as

$$\mathbf{b} = (X'X)^{-1}X'\mathbf{y}, \tag{2.2.5}$$

where $\mathbf{b} = (b_0, b_1, \ldots, b_p)'$, $\mathbf{y} = (y_1, \ldots, y_n)'$ and X is the n by $(p+1)$ matrix with (i, j)th element t_i^{j-1}. When p is large, the numerical stability of the computations can be improved by re-expressing eqn. (2.2.4) in the equivalent form

$$s(t) = \sum_{j=0}^{p} B_j T^j,$$

where $T = t - \bar{t}$ and $\bar{t} = (\sum_{i=1}^{n} t_i)/n$. A further refinement is to replace the powers of t by orthogonal polynomials, which have the property that the resulting matrix $(X'X)$ in eqn. (2.2.5) is diagonal. This use of orthogonal polynomials was essential in the pre-computer age, but its importance is now somewhat diminished.

The polynomial regression approach to smoothing has two superficially attractive features: it copes readily with unequally spaced data and the coefficients b_j are determined automatically by eqn. (2.2.5), leaving only the degree, p, of the fitted polynomial to be determined. In practice, the class of polynomial functions is too restrictive. to provide a generally satisfactory solution to the smoothing problem. A low value of p, say $p = 1$ or 2, gives a useful way of removing simple trends from time-series data. On the other hand, large values of p are seldom useful. Although they give greater flexibility in principle, they impose *global* assumptions about the nature of the data which are seldom warranted, and often produce artefacts.

Example 2.2. Wool prices at weekly markets
The wool-price data of Example 1.3 illustrate this point. Figure 2.5 shows the log-price-ratio, $y_t = \log(19\mu m \text{ price/floor price})$, along with the best-fitting (in the sense of eqn. (2.2.5)) polynomials of degrees $p = 1$ and $p = 10$. The linear fit at least gives a precise, quantitative expression of the overall rising trend during the period July 1976 to June 1984, whilst necessarily failing to describe the more subtle, non-linear fluctuations in the log-price-ratio. The higher-order fit, on the other hand, has a number of unsatisfactory features including a trough in the second half of 1983 where the data suggest a peak, a sharp downturn near the end of the series and, perhaps most noticeably, a displacement by some months of the major peak in 1980.

2.2.3 *Spline regression and its relationship to moving average smoothing*

To motivate spline regression, it is again helpful to think of smoothing as an attempt to estimate the trend, $\mu(t)$, in the basic model

$$Y(t) = \mu(t) + U(t),$$

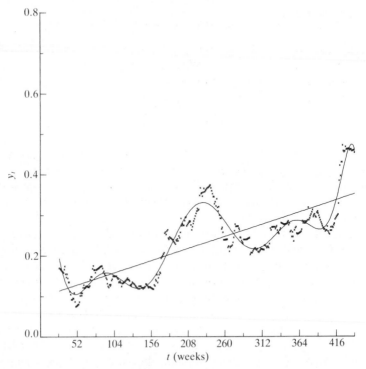

Fig. 2.5. The log-ratio, $y_t = \log(19\,\mu\text{m price/floor price})$, derived from the wool price data. The data are shown as crosses. The solid lines are linear and 10th degree polynomials fitted to the data by ordinary least squares.

where $U(t)$ is a stationary random process. In Section 2.2.2, we constrained $\mu(t)$ to be a polynomial of specified degree and estimated its coefficients by least squares, i.e. for data y_i observed at times t_i, $i = 1, \ldots, n$ we choose the coefficients of the polynomial $\mu(t)$ to minimize the 'residual sum of squares',

$$\sum_{i=1}^{n} \{y_i - \mu(t_i)\}^2.$$

If we now allow $\mu(t)$ to be an arbitrary function of t, this residual sum of squares can be reduced to zero by choosing $\mu(t)$ to be any function which interpolates the data, i.e. one for which $\mu(t_i) = y_i : i = 1, \ldots, n$. There are many such functions, but they are usually far from smooth. For example, we could interpolate using a polynomial of degree $(n - 1)$, by joining the points (t_i, y_i) by straight lines as in most of our time plots, or even by drawing a freehand curve to pass through the points. To counteract this, suppose that we penalize each function $\mu(t)$ by a measure of its roughness. A sensible measure of roughness is the integrated squared second derivative, which is large when the local gradient of $\mu(t)$ changes rapidly, i.e. when $\mu(t)$ is 'wiggly'. This suggests choosing $\mu(t)$ to minimize the quantity

$$Q(\alpha) = \sum_{i=1}^{n} \{y_i - \mu(t_i)\} + \alpha \int_{-\infty}^{\infty} \{\mu''(t)\}^2 \, dt. \qquad (2.2.6)$$

In eqn. (2.2.6), α represents a trade-off between closeness of fit to the data, as measured by the residual sum of squares, and smoothness of $\mu(t)$, as measured by the integral term. If α is close to zero, we tolerate a lot of roughness in $\mu(t)$ in order to fit the data closely, whereas if α is large, we insist on a relatively smooth $\mu(t)$ and allow a less close fit to the data.

For given α, the function $\hat{\mu}(t)$ which minimizes $Q(\alpha)$ is a *cubic spline* (Reinsch 1967). Specifically,

- $\hat{\mu}(t)$ has a continuous first derivative everywhere,
- $\hat{\mu}(t)$ is linear for $t < t_1$ and for $t > t_n$,
- $\hat{\mu}(t)$ is a cubic function of t between each successive pair of t_i.

This is a mathematically elegant result which is at first sight somewhat remote from the data. However, Silverman (1984a) has shown that $\hat{\mu}(t)$ is closely related to a moving average smoother, in the sense that

$$\hat{\mu}(t_i) = \sum_{j=1}^{n} y_j w_{ij}, \qquad (2.2.7)$$

for certain weights w_{ij} which we now define.

Firstly, suppose that the data are equally spaced at unit time intervals. Then,

$$w_{ij} \simeq h^{-1} K\{(i-j)/h\} \qquad (2.2.8)$$

where $h = \alpha^{0.25}$ and the *kernel function* $K(\cdot)$ is defined by

$$K(u) = 0.5 \exp(-|u|/\sqrt{2}) \sin(0.25\pi + |u|/\sqrt{2}) \qquad (2.2.9)$$

(see Fig. 2.6). This shows how an increase in the value of α, corresponding to an increased penalty on the roughness of $\hat{\mu}(t)$, can also be interpreted directly as an increase in the order of the weighted moving average defined by eqns (2.2.7), (2.2.8) and (2.2.9).

More generally, for unequally spaced data eqns (2.2.7) and (2.2.8) generalize to

$$\hat{\mu}(s) = \sum_{j=1}^{n} y_j W(s, t_j) \qquad (2.2.10)$$

where

$$W(s, t) \simeq \{nf(t)h(t)\}^{-1} K\{(s-t)/h(t)\}, \qquad (2.2.11)$$

$$h(t) = \alpha^{0.25}\{nf(t)\}^{-0.25} \qquad (2.2.12)$$

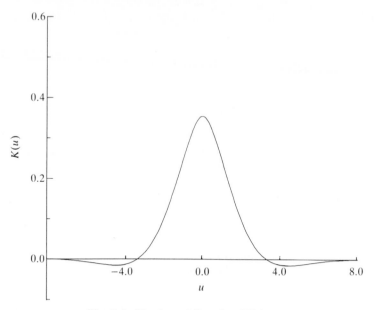

Fig. 2.6. The kernel function $K(u)$.

and $f(t)$ describes the local density of the t_i in a sense which can be made mathematically precise as $n \to \infty$. The important point of eqns (2.2.11) and (2.2.12) is that they show how spline smoothing can be interpreted as a *locally* weighted moving average in which the amount of local smoothing, as measured by $h(t)$, adjusts in a sensible way to variations in the local density of the data by smoothing more in regions where the data are sparse.

A further apparent advantage of spline smoothing is that an automatic method is available for choosing the value of α. This method, known as generalized cross-validation, is due to Craven and Wahba (1979), and efficient algorithms for its implementation have been devised by Hutchinson and de Hoog (1985) and by Silverman (1984b). However, in its standard form, the generalized cross-validation method assumes that the data follow a smooth trend $\mu(t)$ with superimposed *white noise* deviations from the trend. Diggle and Hutchinson (1989) show that this method can give very misleading results when the deviations about the trend are serially dependent. See also Hart and Wehrly (1986).

Computational algorithms for spline smoothing are widely available, for example in the IMSL and NAG subroutine librariees.

Example 2.3. Wool prices at weekly markets.
Figure 2.7 shows the result of applying two different smoothing splines to the wool-price series, $y_t = \log(19 \ \mu\text{m price/floor price})$. In Fig. 2.7a α has

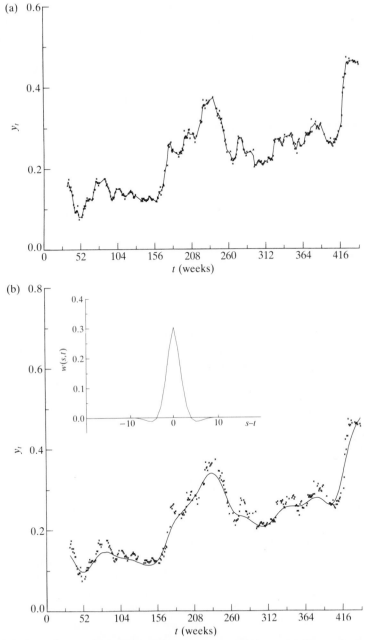

Fig. 2.7. Two smoothing splines fitted to the log-ratio, $y_t = \log(19\,\mu\text{m price/floor}$ price), derived from the wool price data. (a) α chosen by generalized cross-validation. (b) α increased by a factor of 4096, corresponding to an eight-fold increase in the spread of the equivalent moving average (shown as an inset).

been chosen by generalized cross-validation. The resulting estimate $\hat{\mu}(t)$ is very rough. It tracks the data too closely because it is forced to ascribe to $\mu(t)$ any *serially dependent* random fluctuations in prices. This undersmoothing is typical of the results produced by the generalized cross-validation prescription in the presence of unrecognized positive serial correlation. For comparison, Fig. 2.7b shows the result when α is increased by a factor of 4096, corresponding to an eight-fold increase in the spread of the equivalent moving average (shown as an inset to Fig. 2.7b), and consequently a much smoother $\hat{\mu}(t)$.

2.2.4   *Summary*

Moving average smoothing is easy to implement and to interpret. The degree of smoothing can sometimes be chosen from physical considerations, for example to eliminate known seasonal effects. For a subjective, interactive choice, a very convenient method is repeated application of three-point moving averages, inspecting intermediate results.

Polynomial regression provides a convenient method of removing simple trends from time-series data, but is not recommended as a general smoothing procedure. A major disadvantage is that it imposes global assumptions about the nature of any underlying trend, and this can produce undesirable artefacts in the smoothed series.

Spline regression provides a method of implementation of weighted moving average smoothing which copes sensibly with arbitrary patterns of missing values in the data or, more generally, with completely irregularly spaced data. If the data can be assumed to consist of *white noise* deviations about a smooth trend, an automatic choice of the amount of smoothing is available. If not, the amount of smoothing can still be related to the equivalent weighted moving average for equally spaced data, via eqns (2.2.8) and (2.2.9), and this may well help to suggest a sensible amount of smoothing.

2.2.5   *Further reading*

The smoothing methods described in this section are all *linear smoothers,* i.e. the smoothed value at time t is a linear combination of the observations, $s_t = \sum_{j=1}^{n} w_{tj} y_j$. Strictly, this statement is true only if the smoothing weights w_{tj} are chosen without reference to the data, but it nevertheless conveys the essence of what the smoothing operation is doing to the data. In particular, it shows that the effect of a single unusually large observation will persist through a whole sequence of smoothed values; in other words, the smoother is sensitive to outliers in the data. During the last decade there has been a substantial amount of research into *non-linear smoothers* which are less sensitive to outliers, following the pioneering work of Tukey (1977) on exploratory data analysis. For example, Cleveland (1979) develops smoothing procedures based on locally weighted outlier-resistant regression, whilst Velleman (1980) describes an approach based on successive application of moving medians, rather than moving averages.

2.3 Differencing

Differencing provides a simple aproach to removing, rather than high-lighting, trends in time-series data.

The *first difference* of a time series $\{y_t\}$, written $\{Dy_t\}$, is defined by the transformation

$$Dy_t = y_t - y_{t-1}.$$

Higher-order differences are defined by repeated application; thus, for example, the second difference $\{D^2y_t\}$ is defined by

$$D^2y_t = D(Dy_t) = Dy_t - Dy_{t-1} = y_t - 2y_{t-1} + y_{t-2}.$$

Suppose that the series $\{y_t\}$ is a deterministic, linear function of t, $y_t = \alpha + \beta t$, then $Dy_t = (\alpha + \beta t) - \{\alpha + \beta(t-1)\} = \beta$, a constant for all t. More generally, if $\{y_t\}$ consists of a kth degree polynomial in t plus a stationary random component, then $\{D^ky_t\}$ is stationary. This links differencing to the idea of removing a polynomial trend which, as we discussed in Section 2.2.2, can also be achieved by polynomial regression. However, differencing has a wider justification in its ability also to remove a rather specialized form of non-stationarity of the kind exhibited in Example 1.8. In that example, we defined a random sequence $\{y_t\}$ with the property that $\{DY_t\}$ was stationary. In Chapter 6 we shall discuss a general class of non-stationary random sequences with the property that, for some positive integer k, $\{D^kY_t\}$ is stationary.

If differencing is used to remove trend in a time series, the usual recommendation is to construct $\{Dy_t\}$, $\{D^2y_t\}$, etc. until visual inspection suggests that the differenced series, $\{D^ky_t\}$, say, is stationary. In practice, $k = 1$ or 2 usually suffices and it is easy to see that these cases correspond effectively to subtracting from the original series $\{y_t\}$ simple moving averages of order 2 and 3 respectively. For,

$$y_t - \tfrac{1}{2}(y_t + y_{t-1}) = \tfrac{1}{2}(y_t - y_{t-1}) = \tfrac{1}{2}Dy_t$$

and

$$y_t - \tfrac{1}{3}(y_{t-1} + y_t + y_{t+1}) = -\tfrac{1}{3}(y_{t-1} - 2y_t + y_{t+1}) = -\tfrac{1}{3}D^2y_{t-1}.$$

Similarly, higher-order differences correspond to subtraction of higher-order, weighted moving averages.

At first sight, repeated differencing would seem to be very similar in spirit to the repeated application of moving averages of order 3, as suggested in Section 2.2.1. In practice, the two operations give a completely different perspective on a time series, since moving averages focus on the smoothness or otherwise of the estimated trend, whereas differencing emphasizes the apparent stationarity, or 'roughness' of the residuals about the trend.

Example 2.4. Wool prices at weekly markets
Figure 2.8a shows the first difference of the wool-price series $y_t =$ log(19 μm price/floor price). In this example, we have treated the series as equally spaced in time when calculating the first differences. The differenced series shows no obvious non-stationarity, except perhaps for an increase in variability towards the end of the series. Figure 2.8b shows the second difference of y_t. The apparent increase in variance between

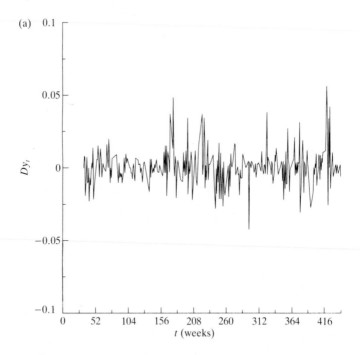

Fig. 2.8. Differencing the log-ratio, $y_t = $ log(19 μm price/floor price), derived from the wool price data. (a) First difference, $Dy_t = y_t - y_{t-1}$. (b) Second difference, $D^2 y_t = y_t - 2y_{t-1} + y_{t-2}$. (c) Moving average, $s_t = \frac{1}{2}(y_t + y_{t-1})$.

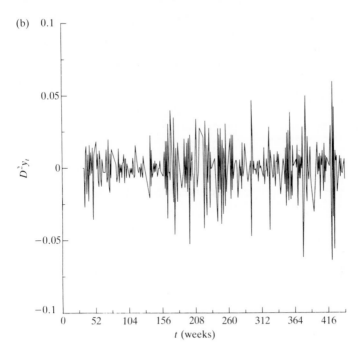

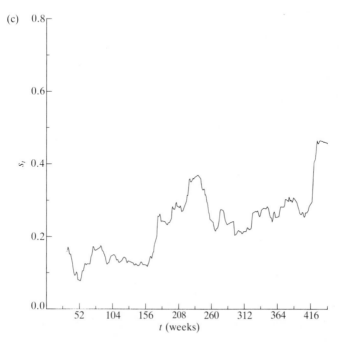

the first and and second differenced series is usually taken as an indication that the data have been over-differenced. On this basis, it would be reasonable to use the series of first differences for further analysis. This corresponds to subtraction from the original series of the two-point moving average shown as Fig. 2.8c. Viewed as an estimate of an underlying trend, Fig. 2.8c is open to the same objection as Fig. 2.7a—it is far too rough. On the other hand if the series $\{y_t\}$ is in fact generated by a stochastic process of the form

$$Y_t = Y_{t-1} + U_t,$$

where $\{U_t\}$ is stationary, then analysis of the first differenced series is exactly the right approach, because there *is* no underlying trend!

2.4 The autocovariance and autocorrelation functions

The *autocovariance function* of a stationary random function $\{Y(t)\}$ is

$$\gamma(k) = \text{cov}\{Y(t), Y(t-k)\}.$$

One consequence of the definition of stationarity is that $\gamma(k)$ does not depend on t. Since $\gamma(0)$ is the variance of each $Y(t)$, we define the *autocorrelation function* as

$$\rho(k) = \gamma(k)/\gamma(0).$$

For a stationary random sequence $\{Y_t\}$, k is necessarily an integer and we shall then usually write γ_k and ρ_k for the *autocovariance* and *autocorrelation coefficients* of $\{Y_t\}$.

The autocorrelation function is an important, albeit incomplete, summary of the serial dependence within a stationary random function. Before giving some examples to illustrate this point, we mention some general properties of $\rho(k)$ which, whilst elementary, are important:

(a) $\rho(k) = \rho(-k)$,
(b) $-1 \le \rho(k) \le 1$,
(c) if $Y(t)$ and $Y(t-k)$ are independent, then $\rho(k) = 0$.

The proof of (a) is immediate, whilst (b) and (c) are standard properties of the correlation between a pair of random variables, and proofs can be found in any elementary book on probability theory. It is worth emphasizing that the converse of (c) is false.

A simple example of an autocorrelation function is the following.

Example 2.5. A first-order autoregressive process
Consider a random sequence $\{Y_t\}$ defined by

$$Y_t = \alpha Y_{t-1} + Z_t, \qquad (2.4.1)$$

where $\{Z_t\}$ is a white noise sequence. In Chapter 3 we shall prove that $\{Y_t\}$ is stationary if and only if $-1 < \alpha < 1$. It follows that, at least for this range of values of α, $E(Y_t) = \mu$ does not depend on t. Taking expectations on both sides of eqn. (2.4.1), and remembering that $E(Z_t) = 0$, we deduce that $\mu = \alpha\mu$, and therefore that $\mu = 0$. Now, multiplying both sides of eqn. (2.4.1) by Y_{t-k}, taking expectations and dividing by $\mathrm{Var}(Y_t)$ gives

$$\rho_k = \alpha\rho_{k-1}.$$

Finally, $\rho_0 = 1$ gives the solution

$$\rho_k = \alpha^k \quad : \quad k = 0, 1, \dots. \tag{2.4.2}$$

Figure 2.9 shows the autocorrelation function (2.4.2) for each of $\alpha = -0.5$, 0.5 and 0.9, along with simulated partial realizations of $\{Y_t\}$ incorporating Normally distributed white noise sequences $\{Z_t\}$. The alternating pattern of autocorrelations in the case $\alpha = -0.5$ reflects a tendency for the series $\{Y_t\}$ to zigzag about zero, whilst the strictly positive autocorrelations in the case $\alpha = 0.5$ impart a degree of smoothness in $\{Y_t\}$. This smoothness is more evident in the final case, $\alpha = 0.9$, for which $\{Y_t\}$ takes relatively long excursions above and below its average value of zero; as we have already seen in Example 1.8, the limiting case as α approaches 1 gives a *non-stationary* random sequence.

The autocorrelation function gives a complete description of the serial dependence within a stationary *Gaussian* process, i.e. one for which the joint distribution of $Y(t_1), \dots, Y(t_n)$ is multivariate Normal for any n and $t_1, \dots, t_n$. The following example illustrates that autocorrelation is in general an *incomplete* descriptor of serial dependence, in the sense that random processes whose realizations are qualitatively different can share the same autocorrelation function.

Example 2.6. A non-Gaussian autoregression
Consider a first-order autoregressive process in which the Normal distribution for Z_t used in example 2.5 is replaced by a two-component mixture,

$$Z_t = \begin{cases} 0 & \text{with probability } p \\ U_t & \text{with probability } 1 - p, \end{cases}$$

where $\{U_t\}$ is a Normally distributed white noise sequence. Clearly, for large p, realizations of this process differ in an obvious way from that shown in Example 2.5 by their inclusion of exponentially decaying subsequences corresponding to successive $Z_t = 0$. But this has no bearing on the autocorrelation structure of $\{Y_t\}$. Figure 2.10 shows partial realizations with $\alpha = 0.5$ and each of $p = 0.5$, 0.8 and 0.9. These all have

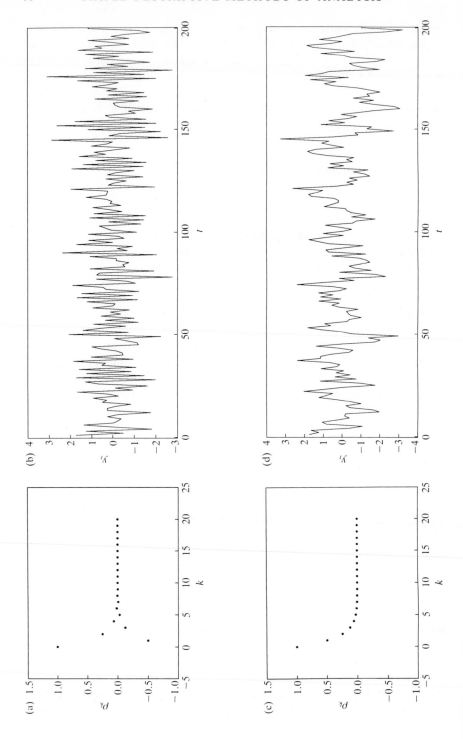

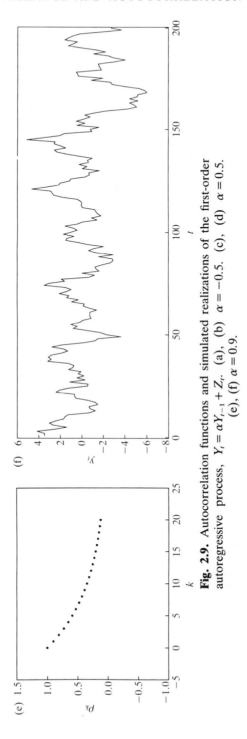

Fig. 2.9. Autocorrelation functions and simulated realizations of the first-order autoregressive process, $Y_t = \alpha Y_{t-1} + Z_t$. (a), (b) $\alpha = -0.5$. (c), (d) $\alpha = 0.5$. (e), (f) $\alpha = 0.9$.

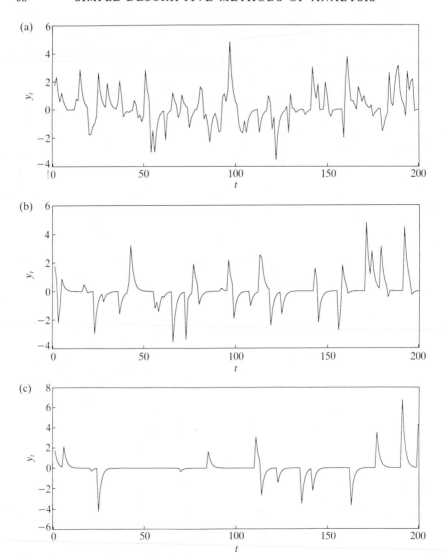

Fig. 2.10. Simulated realizations of a first-order autoregressive process, $Y_t = 0.5Y_{t-1} + Z_t$, in which each $Z_t = 0$ with probability p. (a) $p = 0.5$. (b) $p = 0.8$. (c) $p = 0.9$.

the same autocorrelation structure but are obviously different in other respects.

2.5 Estimating the autocorrelation function

Estimation of an autocorrelation function from a set of data proceeds along rather different lines for series observed at equally or unequally spaced times, and we shall treat the two cases separately.

2.5.1 *Equally spaced series: the correlogram*

For a series $\{y_t : t = 1, \ldots, n\}$ we use $\bar{y}$ to denote the sample mean, $\bar{y} = (\sum y_i)/n$, and define the kth *sample autocovariance coefficient*,

$$g_k = \sum_{t=k+1}^{n} (y_t - \bar{y})(y_{t-k} - \bar{y})/n. \qquad (2.5.1)$$

Then, the kth *sample autocorrelation coefficient* is

$$r_k = g_k/g_0. \qquad (2.5.2)$$

A plot of r_k against k is called the *correlogram* of the data $\{y_t\}$. It might seem more natural to define the g_k with divisors $n - k$, but the definition given here is the conventional one. From a practical point of view, the distinction is unimportant as long as k is very much smaller than n; this is seldom a severe restriction since values of r_k for large k are difficult to interpret.

One simple use of the correlogram is to check whether there is evidence of any serial dependence in an observed time series. To do this, we use a result due to Bartlett (1946), who showed that for a white noise sequence $\{y_t\}$ and for large n, r_k is approximately Normally distributed with mean zero and variance $1/n$. We write this as

$$r_k \sim N(0, 1/n). \qquad (2.5.3)$$

Thus, values of r_k greater than $2/\sqrt{n}$ in absolute value can be regarded as significant at about the 5% level. If a large number of r_k are calculated it is likely that some will exceed this threshold even if $\{y_t\}$ is a white noise sequence. Interpretation is further complicated by the fact that the r_k are themselves statistically dependent, i.e. the probability of any given r_k being greater than $2/\sqrt{n}$ in absolute value depends on the observed values of other r_k. Box and Pierce (1970) developed a 'portmanteau' test of white noise later refined by Ljung and Box (1978), who showed that, for a white noise sequence $\{y_t : t = 1, \ldots, n\}$, large n and $m \ll n$,

$$Q_m = n(n+2) \sum_{k=1}^{m} (n-k)^{-1} r_k^2 \sim \chi_m^2 \tag{2.5.4}$$

The sensitivity of the Q_m-statistic to various types of departure from white noise depends on the choice of m. For some further comments, see Section 2.8.

In many applications, the white noise hypothesis is untenable, and a more constructive use of the correlogram is to shed some light on the nature of the serial dependence within a set of data. For this purpose, the overall pattern of the correlogram is more important than the precise numerical values of individual r_k. We can then relate the qualitative behaviour of the correlogram to various theoretical forms of the autocorrelation function with a view to suggesting plausible models. For example, a correlogram which appears to decay exponentially would suggest a first-order autoregressive process as a possible model. Obviously, the reliability of the correlogram for this purpose increases with the length of the time-series on which it is based.

Example 2.7. Correlograms of simulated first-order autoregressive sequences

Figure 2.11 shows the correlograms of the five simulated series shown in Figures 1.10, 1.11 and 2.9d,e,f. These are all of length $n = 200$, and have been generated by the first-order autoregressive process,

$$Y_t = \alpha Y_{t-1} + Z_t,$$

where $\{Z_t\}$ is Gaussian white noise. For $-1 < \alpha < 1$, the autocorrelation function of $\{Y_t\}$ is

$$\rho_k = \alpha^k \quad : \quad k = 0, 1, 2, \ldots .$$

These correlograms correspond to

(a) $\alpha = 0$ (c.f. Fig. 1.10)
(b) $\alpha = -0.5$ (c.f. Fig. 2.9d)
(c) $\alpha = 0.5$ (c.f. Fig. 2.9e)
(d) $\alpha = 0.9$ (c.f. Fig. 2.9f)
(e) $\alpha = 1.0$ (c.f. Fig. 1.11)

Each correlogram includes a pair of dashed horizontal lines representing the limits $\pm 2/\sqrt{n}$, which are used for informal assessment of departure from white noise. For all except case (a), the correlograms clearly, and correctly, suggest significant departure from white noise. Furthermore, at least for small k, the correlograms in cases (b), (c) and (d) show good agreement with the corresponding theoretical autocorrelation functions

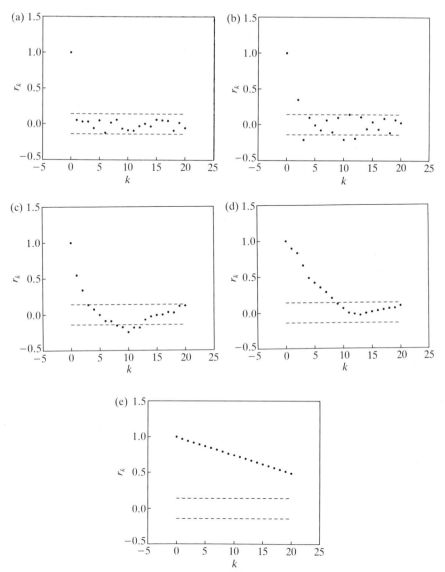

Fig. 2.11. Correlograms of simulated realizations of a first-order autoregressive process $Y_t = \alpha Y_{t-1} + Z_t$. The dashed horizontal lines on each correlogram represent the limits $\pm 2/\sqrt{n}$ (a) $\alpha = 0.0$. (b) $\alpha = -0.5$. (c) $\alpha = 0.5$. (d) $\alpha = 0.9$. (e) $\alpha = 1.0$.

shown in Fig. 2.9a,b,c. At larger lags, sampling fluctuations tend to hide the underlying pattern in each correlogram, since the theoretical autocorrelations are close to zero. Note in particular the spurious trough in case (c) around $k = 10$. Finally, the slow, approximately linear decay in case (e) is typical of the behaviour of the correlogram for a non-stationary time series whose theoretical autocorrelation function does not exist.

Example 2.8. Correlograms of some real data-sets

(a) Levels of luteinizing hormone (LH) in blood samples.

Figure 2.12 shows the correlograms of the three LH series from Example 1.2, along with the $\pm 2/\sqrt{n}$ limits for assessing departure from white noise. In all three cases, the first autocorrelation coefficient is significantly positive. Additonally, an interesting feature of the two late follicular series (Figs. 2.12a and 2.12c) is the cyclic pattern of autocorrelations, suggesting a similarly cyclic pattern of time variation in LH. In Section

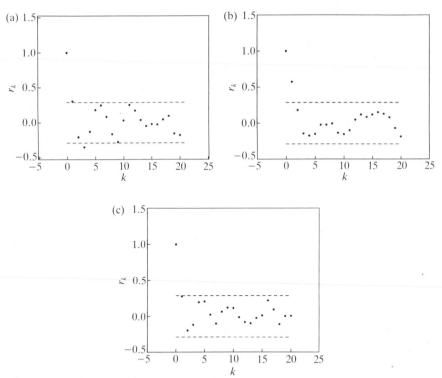

Fig. 2.12. Correlograms of three LH series. The dashed horizontal lines on each correlogram represent the limits $\pm 2/\sqrt{n}$. (a) Late follicular phase of first menstrual cycle. (b) Early follicular phase of second menstrual cycle. (c) Late follicular phase of second menstrual cycle.

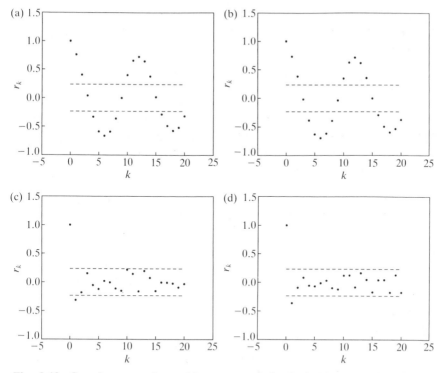

Fig. 2.13. Correlograms of monthly returns of deaths in the United Kingdom attributed to bronchitis, emphysema, and asthma. (a) Males. (b) Females. (c) Male residuals. (d) Female residuals.

2.7 we shall define another statistical tool, the periodogram, which examines cyclic patterns more directly.

(b) UK deaths from bronchitis, emphysema, and asthma.

Figures 2.13a,b show the correlograms of the male and female series from Example 1.4. Both correlograms clearly reflect the strong seasonal pattern in the data, with troughs at lags 6 and 18 and a peak at lag 12. One way to remove the seasonal pattern in the data is to subtract from each series, y_t, the three-point moving average, s_t, shown in Fig. 2.4a. The correlograms of the two residual series, $y_t - s_t$, are shown in Figs. 2.13c,d. Apart from a negative autocorrelation at lag one, both correlations lie entirely within the $\pm 2/\sqrt{n}$ limits. In Section 2.6 we shall show that the negative correlation at lag one in the residual series can be explained by our having used a three-point moving average to construct the smooth series s_t. Thus, a provisional description of these data might be that they consist of a strong seasonal trend with independent random perturbations from month to month. In Section 4.7 we shall subject the data to more careful scrutiny to see whether this simple interpretation is sustainable.

2.5.2   Unequally spaced series: the variogram

For a stationary random process $\{Y(t)\}$, the distribution of $Y(t) - Y(t - k)$ does not depend on t. In particular, if $\gamma(k)$ denotes the autocovariance function of $\{Y(t)\}$ and $\rho(k) = \gamma(k)/\gamma(0)$ the corresponding autocorrelation function, then we can define the *variogram*, $V(k)$, as

$$V(k) = \tfrac{1}{2}E[\{Y(t) - Y(t - k)\}^2] = \gamma(0)\{1 - \rho(k)\}. \qquad (2.5.5)$$

This result provides the key to estimation of $\rho(k)$ from data observed at unequally spaced times. For a series $\{y(t_i) : i = 1, \ldots, n\}$, a plot of the quantities $v_{ij} = \tfrac{1}{2}\{y(t_i) - y(t_j)\}^2$ against $k_{ij} = t_i - t_j$ for all $\tfrac{1}{2}n(n - 1)$ distinct pairs of observations is called the *empirical* or *sample variogram*. The terminology is due to the French geostatistical school, who use the concept extensively in the body of techniques known as kriging; see, for example, Journel and Huijbregts (1978). Note that in the geostatistical literature, $V(u)$ is called the semivariogram because of the factor $\tfrac{1}{2}$ in eqn. (2.5.5).

Unless the times of observations are completely irregular, there will be more than one v_{ij} corresponding to a particular value k of k_{ij}. Replacing all such v_{ij} by their mean value, $\bar{v}(k)$, leads to a desirable reduction in the amount of random scatter in the sample variogram. Typically, the $\bar{v}(k)$ tend to fluctuate around a constant value as k increases because the autocorrelations decay to zero. Then, eqn. (2.5.5) suggests that we can estimate this limiting value by

$$v_\infty = g_0 = \sum_{t=1}^{n} (y_t - \bar{y})^2/n,$$

to give an estimate of $\rho(k)$ as

$$\hat{\rho}(k) = 1 - \bar{v}(k)/v_\infty.$$

If $\bar{v}(k)$ shows no sign of levelling off within the available range of k, the implication is either that the serial dependence in $\{Y(t)\}$ extends beyond the time-span covered by the data, or that the underlying process is non-stationary. Note that the function $V(k)$ exists for some non-stationary processes. For example, the random walk process defined in Example 1.8 has $V(k) = \tfrac{1}{2}k\sigma^2$.

Example 2.9. The sample variogram of an LH series with random deletions

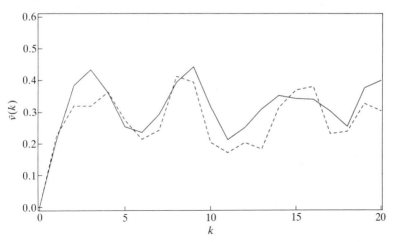

Fig. 2.14. Sample variograms of the first of the three LH time series. (a) ———, Calculated from all 48 observations. (b) – – –, calculated after random deletion of 15 out of 48 observations.

Figure 2.14 shows the sample variogram of the first of the three LH series in Example 1.2. Comparison with Fig. 2.12a shows a similar, cyclic pattern to both the correlogram and the sample variogram of these data. Since our main motivation for introducing the variogram is as a means of coping with unequally spaced data, we also show in Fig. 2.14 the sample variogram obtained after randomly deleting 15 of the 48 LH values. Although there are non-negligible changes in the individual $\bar{v}(k)$, the overall cyclic pattern persists and the qualitative conclusions about the data are unchanged.

2.6 Impact of trend removal on autocorrelation structure

Recall our basic working model for a non-stationary random function,

$$Y(t) = \mu(t) + U(t),$$

where $\{U(t)\}$ is stationary. If we knew $\mu(t)$ exactly, we could subtract it from an observed series $y(t_i)$ to obtain a stationary series $\{u(t_i) : i = 1, \ldots, n\}$ which would be a partial realization of $\{U(t)\}$. In practice, $\mu(t)$ is invariably unknown, and by subtracting an estimate $\hat{\mu}(t) \neq \mu(t)$ we obtain a 'corrupted' *residual series* $\{r(t_i) : i = 1, \ldots, n\}$, where $r(t) = y(t) - \hat{\mu}(t) \neq u(t)$. This corruption has implications for how we make inferences about $U(t)$ from analysis of the series $\{r(t_i)\}$.

For clarity, we shall discuss only the simple case in which the data consist of an equally spaced series $\{y_t : t = 1, \ldots, n\}$, the $\{u_t : t = 1, \ldots, n\}$ form a white noise sequence, and $\hat{\mu}_t$ is an unweighted moving average of order $2p + 1$. Furthermore, we shall assume initially that μ_t is a linear function of t. Then,

$$r_t = y_t - \hat{\mu}_t = \left\{ \mu_t - (2p+1)^{-1} \sum_{j=-p}^{p} \mu_{t+j} \right\} + \left\{ u_t - (2p+1)^{-1} \sum_{j=-p}^{p} u_{t+j} \right\}$$
$$= d_t + (u_t - e_t), \tag{2.6.1}$$

say. However, the linear form of μ_t implies that $\mu_{t+j} + \mu_{t-j} = 2\mu_t$ and $d_t = 0$. Ideally, we should like r_t to be approximately equal to u_t. Now, u_t is a realization of a random variable with zero mean and variance σ^2, whereas the 'corruption',

$$e_t = (2p+1)^{-1} \sum_{j=-p}^{p} u_{t+j}$$

has zero mean and variance $\sigma^2/(2p+1)$. Apparently, to minimize the corruption we should take p as large as possible. This is not a useful suggestion for at least two reasons. Firstly, if p increases without limit we are unable to calculate r_t except at times increasingly close to the middle of the total time interval over which observations are made. Secondly, the essential feature of moving average smoothing is that it avoids the necessity to assume a simple parametric form for μ_t. In general, we need to compromise between keeping d_t in eqn. (2.6.1) small by taking p small, and keeping e_t small by taking p large. Qualitatively, the same argument holds when $\{U_t\}$ is an autocorrelated sequence.

One specific consequence of trend-removal is that it induces spurious autocorrelation into the residual sequence $\{r_t\}$. For example, if we take $p = 1$ and assume that this effectively gives $d_t = 0$ to a good approximation, then

$$r_t = u_t - (u_{t-1} + u_t + u_{t+1})/3$$
$$= (-u_{t-1} + 2u_t - u_{t+1})/3.$$

Now, consider the stationary random sequence $R_t = -\frac{1}{3}(U_{t-1} - 2U_t + U_{t+1})$, and let γ_k denote its autocovariance function. Then,

$$\gamma_0 = E[(U_{t-1} - 2U_t + U_{t+1})^2]/9 = 2\sigma^2/3,$$

where $\sigma^2 = E[U_t^2]$, because $E[U_t U_s] = 0$ unless $t = s$. Similarly,

$$\gamma_1 = E[(U_{t-1} - 2U_t + U_{t+1})(U_{t-2} - 2U_{t-1} + U_t)]/9$$
$$= E[-2U_{t-1}^2 - 2U_t^2]/9 = -4\sigma^2/9,$$
$$\gamma_2 = E[(U_{t-1} - 2U_t + U_{t+1})(U_{t-3} - 2U_{t-2} + U_{t-1})]/9$$
$$= E[U_{t-1}^2]/9 = \sigma^2/9$$

and $\gamma_k = 0$ for $k \geq 3$. Thus, the induced autocorrelation function of $\{R_t\}$ is

$$\rho_k = \begin{cases} 1 & : \quad k = 0 \\ -\frac{2}{3} & : \quad k = 1 \\ \frac{1}{6} & : \quad k = 2 \\ 0 & : \quad k \geq 3 \end{cases}.$$

This discussion becomes particularly relevant if low-order differencing is used to remove smooth trends in non-stationary time series.

A corresponding analysis of the effect of smoothing using general moving averages would involve similarly elementary, if slightly messy, algebraic manipulations. The number of non-zero induced autocorrelations is equal to one less than the order of the moving average, but their numerical magnitudes decrease as the order increases. For example, using a $(2p + 1)$-point unweighted moving average, the induced autocorrelation at lag 1 is

$$\rho_1 = -(p + 1)/\{p(2p + 1\} \simeq -1/(2p).$$

This phenomenon is related to the properties of a class of random processes known as linear filters, which we shall discuss in Chapter 3.

2.7 The periodogram

The periodogram is a summary description based on a representation of an observed time series as a superposition of sinusoidal waves of various frequencies. The practical value of this approach to time-series analysis rests most obviously on the empirical observation that many time series exhibit cyclic fluctuations in value, but at frequencies which are not always predictable before the data are examined.

To motivate the periodogram, we shall examine a very simple model for a time series $\{y_t\}$ exhibiting cyclic fluctuations with a known period, p, say. We assume that

$$y_t = \alpha \cos(\omega t) + \beta \sin(\omega t) + z_t \quad : \quad t = 1, \ldots, n \qquad (2.7.1)$$

where $\{z_t\}$ is a white noise sequence, $\omega = 2\pi/p$ is the known *frequency* of the cyclic fluctuations, and α, β are parameters which we shall estimate by least squares. Write $\mathbf{y} = (y_1, \ldots, y_n)'$, $\boldsymbol{\theta} = (\alpha, \beta)'$, $c_t = \cos(\omega t)$ $\mathbf{c} = (c_1, \ldots, c_n)'$, $s_t = \sin(\omega t)$, and $\mathbf{s} = (s_1, \ldots, s_n)'$. Also, define the n by 2 matrix X to have columns $\mathbf{c}$ and $\mathbf{s}$. Then (Appendix B), the least squares estimate of $\boldsymbol{\theta}$ is

$$\hat{\boldsymbol{\theta}} = (X'X)^{-1}X'\mathbf{y}. \qquad (2.7.2)$$

Before examining (2.7.2) in detail, we remark that in practice we shall

consider only frequencies in the range 0 to π, the upper limit cor-
respondng to an alternating sequence of positive and negative values. For
further comment, see Section 3.2. We shall further restrict ω to be of the
form $\omega = 2\pi j/n$, for some positive integer $j < n/2$. This restriction
simplifies the algebra and has other desirable consequences which will
emerge in due course. Note that the practical force of the restriction to
this discrete set of frequencies, called the *Fourier frequencies*, is
negligible if n is large, but not otherwise. The algebraic simplification
stems from the following result.

Theorem 2.1. Let $\omega_j = 2\pi j/n$ for some positive integer $j < n/2$, $c_t = \cos(\omega t)$ and $s_t = \sin(\omega t)$. Then,

(i)
$$\sum_{t=1}^{n} c_t = \sum_{t=1}^{n} s_t = \sum_{t=1}^{n} c_t s_t = 0,$$

(ii)
$$\sum_{t=1}^{n} c_t^2 = \sum_{t=1}^{n} s_t^2 = n/2.$$

Proof. The proof uses the complex number representation $e^{i\theta} = \cos\theta + i\sin\theta$ (see Appendix C). Write

$$\sum_{t=1}^{n} e^{i\omega t} = \{e^{i\omega}(1 - e^{i\omega n})\}/(1 - e^{i\omega}) = 0 \qquad (2.7.3)$$

for ω any integer multiple of $2\pi/n$. This follows because $e^{2\pi i} = \cos(2\pi) + i\sin(2\pi) = 1$. Re-write eqn. (2.7.3) as

$$\left\{\sum_{t=1}^{n} \cos(\omega t)\right\} + i\left\{\sum_{t=1}^{n} \sin(\omega t)\right\} = 0 + 0i. \qquad (2.7.4)$$

Equating real and imaginary parts in eqn. (2.7.4) gives $\sum c_t = \sum s_t = 0$.
Now, write

$$\sum_{t=1}^{n} c_t s_t = \tfrac{1}{2}\sum_{t=1}^{n} \sin(2\omega t) = 0,$$

by the same argument. Finally, write

$$\sum_{t=1}^{n} c_t^2 = \tfrac{1}{2}\sum_{t=1}^{n} \{1 + \cos(2\omega t)\} = \tfrac{1}{2}\left\{n + \sum_{t=1}^{n} \cos(2\omega t)\right\} = n/2$$

and

$$\sum_{t=1}^{n} s_t^2 = \tfrac{1}{2}\sum_{t=1}^{n} \{1 - \cos(2\omega t)\} = \tfrac{1}{2}\left\{n - \sum_{t=1}^{n} \cos(2\omega t)\right\} = n/2.$$

This completes the proof. ∎

Returning now to eqn. (2.7.1), Theorem (2.1) implies that

$$X'X = \begin{bmatrix} \sum_{t=1}^{n} c_t^2 & \sum_{t=1}^{n} c_t s_t \\ \sum_{t=1}^{n} c_t s_t & \sum s_t^2 \end{bmatrix} = \begin{bmatrix} n/2 & 0 \\ 0 & n/2 \end{bmatrix}.$$

Writing the column vector $X'\mathbf{y}$ as

$$X'\mathbf{y} = \begin{bmatrix} \sum_{t=1}^{n} y_t c_t \\ \sum_{t=1}^{n} y_t s_t \end{bmatrix}$$

we obtain the explicit result

$$\hat{\alpha} = 2\left\{ \sum_{t=1}^{n} y_t \cos(t\omega) \right\} \Big/ n,$$

$$\hat{\beta} = 2\left\{ \sum_{t=1}^{n} y_t \sin(t\omega) \right\} \Big/ n.$$

Again appealing to results in Appendix B, we can write the regression sum of squares for the fitted sinusoid as

$$\mathbf{y}'X(X'X)^{-1}X'\mathbf{y} = n(\hat{\alpha}^2 + \hat{\beta}^2)/2$$
$$= 2\left[\left\{ \sum y_t \cos(\omega t) \right\}^2 + \left\{ \sum y_t \sin(\omega t) \right\}^2 \right] \Big/ n. \quad (2.7.5)$$

Note that this is proportional to the squared amplitude, $\hat{\alpha}^2 + \hat{\beta}^2$, of the fitted sinusoid. Note also that, in the terminology of regression analysis, the regression sum of squares is on two degrees of freedom, one each for α and for β.

Now consider a slightly less simplistic model in which we contemplate not one, but several, sinusoidal components. This gives a model of the form

$$y_t = \sum_{k=1}^{m} \{ \alpha_k \cos(\omega_k t) + \beta_k \sin(\omega_k t) \} + u_t \quad : \quad t = 1, \ldots, n, \quad (2.7.6)$$

where $\{U_t\}$ is again a white noise sequence and each ω_k is one of the Fourier frequencies. Estimation of the parameters α_k and β_k by least squares again gives a simple, explicit result because of the following corollary to Theorem 2.1.

Corollary *Let* $\omega_1 = 2\pi j_1/n$ *and* $\omega_2 = 2\pi j_2/n$ *for distinct positive integers*

$j_1 < n/2$ and $j_2 < n/2$. Write $c_{kt} = \cos(\omega_k t)$ and $s_{kt} = \sin(\omega_k t)$. Then,

$$\sum_{t=1}^{n} c_{1t}c_{2t} = \sum_{t=1}^{n} s_{1t}s_{2t} = \sum_{t=1}^{n} c_{1t}s_{2t} = 0.$$

Proof. The proof is essentially the same as for Theorem 2.1, noting that

$$c_{1t}c_{2t} = \tfrac{1}{2}[\cos\{(\omega_1 + \omega_2)t\} + \cos\{(\omega_1 - \omega_2)t\}],$$
$$s_{1t}s_{2t} = -\tfrac{1}{2}[\cos\{(\omega_1 + \omega_2)t\} - \cos\{(\omega_1 - \omega_2)t\}]$$

and

$$c_{1t}s_{2t} = \tfrac{1}{2}[\sin\{(\omega_1 + \omega_2)t\} - \sin\{(\omega_1 - \omega_2)t\}]. \qquad \blacksquare$$

The above corollary implies that in the matrix representation of eqn. (2.7.6), $X'X$ is again a multiple of the identity matrix and $\hat{\alpha}_k$, $\hat{\beta}_k$ can be written explicitly as

$$\hat{\alpha}_k = 2\left\{\sum_{t=1}^{n} y_t \cos(\omega_k t)\right\}\Big/n,$$

$$\hat{\beta}_k = 2\left\{\sum_{t=1}^{n} y_t \sin(\omega_k t)\right\}\Big/n.$$

Furthermore, expression (2.7.5) again gives the sums of squares associated with each of the m fitted sinusoids. Finally, the diagonal nature of $X'X$ implies that the regression sum of squares associated with the model (2.7.6) is precisely the sum of the m separate sums of squares associated with each sinusoidal component.

It should now be clear that by increasing m in eqn. (2.7.6) we can achieve an orthogonal partitioning of progressively more of the variation in the series $\{y_t\}$ into sinusoidal components, each with two degrees of freedom. In particular, if we take m to be the largest integer less than $n/2$ and $\omega_k = 2\pi k/n$ we obtain a regression sum of squares with degrees of freedom $2m$, equal to $n - 1$ if n is odd and $n - 2$ if n is even. To achieve a complete partitioning, we add the extreme frequencies $\omega = 0$ and (if n is even) $\omega = \pi$. Consider first $\omega = 0$. The sine terms now vanish and the cosines are identically 1, so our original model (2.7.1) reduces to

$$y_t = \alpha + z_t \quad : \quad t = 1, \ldots, n.$$

Clearly $\hat{\alpha} = \bar{y}$, the sample mean, and the associated contribution to the regression sum of squares is $n\bar{y}^2$ on one degree of freedom. Similarly, for $\omega = \pi$ and n even, eqn. (2.7.1) reduces to

$$y_t = \alpha(-1)^t + z_t \quad : \quad t = 1, \ldots, n,$$

$\hat{\alpha} = \{\sum y_t(-1)^t\}/n$, and the associated sum of squares, again on one degree of freedom, is $\{\sum_{t=1}^{n} y_t(-1)^t\}^2/n$.

We are now in a position to define the *periodogram ordinates* $I(\omega)$, for $0 \leqslant \omega \leqslant \pi$, by

$$I(\omega) = \left[\left\{ \sum_{t=1}^{n} y_t \cos(\omega t) \right\}^2 + \left\{ \sum_{t=1}^{n} y_t \sin(\omega t) \right\}^2 \right] \Big/ n. \qquad (2.7.7)$$

A comparison between eqns (2.7.5) and (2.7.7) shows that they differ by a factor of 2. The reason for this apparent perversity will emerge soon. For the time being, note that we can express our partitioning of the total variation in the series $\{y_t\}$ as

$$\sum_{t=1}^{n} y_t^2 = I(0) + 2 \sum_{j=1}^{m} I(2\pi j/n) + I(\pi),$$

where m is the largest integer less than $n/2$, and the final term $I(\pi)$ is absent if n is odd.

A plot of $I(\omega)$ as ordinate against ω as abscissa is called the periodogram of $\{y_t\}$. Although $I(\omega)$ is defined for all ω in the range 0 to π, it is usual to plot only ordinates associated with positive Fourier frequencies. The reason for excluding $\omega = 0$ is that $I(0)$, corresponding as it does to the sample mean $\bar{y}$, is of limited interest in the present context of exploring cyclic patterns in time-series data. With regard to non-Fourier frequencies ω, eqn. (2.7.7) should then strictly be regarded as a convenient approximation, since it no longer represents precisely the regression sum of squares associated with a sinusoid of frequency ω. Furthermore, the statistical properties of the periodogram are greatly simplified by the restriction to Fourier frequencies as a direct consequence of the orthogonality of the corresponding partitioning of the total sum of squares $\sum_{t=1}^{n} y_t^2$.

Example 2.10. Periodograms of some real data sets

(a) Levels of luteinizing hormone (LH) in blood samples

Figure 2.15 shows the periodogram of the first of the three LH series from Example 1.2. The dominant peak indicates a cyclic component to the time variation in the data, the location of the peak suggesting a frequency of about one cycle per hour, i.e. 8 complete cycles in the 8-hour time-span of the data. However, the non-negligible contributions to the periodogram from other frequencies indicate that a simple sinusoid at a frequency of one cycle per hour would leave unexplained a substantial proportion of the variation in the data.

(b) UK deaths due to bronchitis, emphysema, and asthma

In Example 2.8 we noted that the data on male and female UK deaths due to bronchitis, emphysema, and asthma (Example 1.4) showed a much stronger cyclic pattern than the LH data. Furthermore, we know *a priori* that the cyclic pattern is seasonal, repeating itself every 12 months.

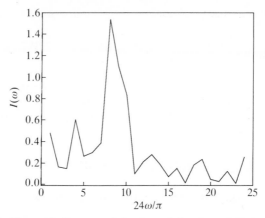

Fig. 2.15. The periodogram of the first of the three LH time series.

Since the data cover a time-span of 72 months, we can therefore expect the periodograms of these two series to be dominated by the contributon from the sixth Fourier frequency, corresponding to the annual cycle. Figure 2.16 confirms this, as well as revealing that for both series the next largest periodogram ordinates are at the twelfth Fourier frequency, corresponding to an apparent six-month cycle. This last comment is slightly misleading – a more accurate statement is that the superposition of two harmonic components gives an improved description of an annual cycle which is not a pure sinusoid. For further discussion, see Section 4.10. Finally, the small but noticeable contribution to the periodogram of the male series at the first Fourier frequency is a reflection of the gentle long-term trend in the data previously demonstrated in Fig. 2.4b.

Examples 2.10a and 2.10b provide an interesting contrast between the relatively straightforward interpretation of the periodogram in Example 2.10b, for which the data were known *a priori* to exhibit strong cyclic variation at one of the Fourier frequencies, and the more subtle pattern of the periodogram in Example 2.10a, where the cyclic variation was weaker and not necessarily at a frequency corresponding to any of the Fourier frequencies. To emphasize the difficulties which can arise in the interpretation of periodograms, Fig. 2.17 shows a simulated realization of white noise and its associated periodogram. For a white noise process, the periodogram ordinates are mutually independent, all with unit mean and sampling distributions proportional to chi-squared on two degrees of freedom. This distribution has a long upper tail, the effect of which is that the periodogram includes many visually striking peaks, despite the

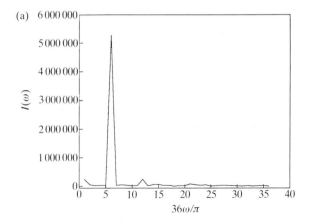

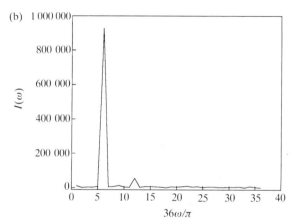

Fig. 2.16. Periodograms of monthly returns of deaths in the United Kingdom attributed to bronchitis, emphysema, and asthma. (a) Males. (b) Females.

complete absence of any genuine cyclic effects in the process which generated the daa.

The above remarks notwithstanding, the periodogram provides the basis for a generally useful approach to the analysis of stationary time series. This approach, spectral analysis, is the subject of Chapter 4. For the time being, we describe one application of the periodogram, an alternative to the Box–Pierce portmanteau statistic (2.5.4) for testing the hypothesis that $\{y_t\}$ is a white noise sequence; a rigorous justification of the procedure will be given in Chapter 4.

Define quantities $C_j = \sum_{k=1}^{j} I(2\pi k/n) : j = 1, \ldots, m$, where m is the largest integer strictly less than $n/2$, and $U_j = C_j/C_m$. A plot of U_j against j/m', where $m' = m - 1$, is called the *cumulative periodogram*. Bartlett

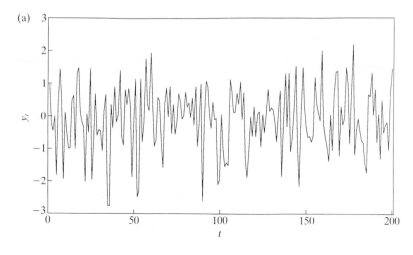

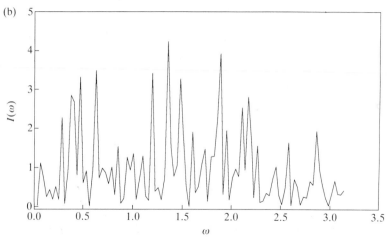

Fig. 2.17. A simulated realization of Normally distributed white noise and the corresponding periodogram. (a) Simulated realization. (b) Periodogram.

(1954, 1955) suggested that the cumulative periodogram might be used to test the hypothesis that $\{y_t\}$ is a white noise sequence. The rationale for this is that under the white noise hypothesis, all the 'true' amplitudes of the sinusoidal components in the model (2.7.6) are zero, the periodogram ordinates should therefore differ only because of sampling fluctuations, and the cumulative periodogram should increase approximately linearly from 0 to 1 as j runs from 1 to m'. A statistic which measures departure from linearlity is the maximum horizontal deviation of the plotted points

$(j/m', U_j)$ from the line $y = x$. In algebraic terms, this is expressible as

$$D = \max_{j=1,\dots,m'} \{\max(|U_j - j/m'|, |U_j - (j-1)/m'|)\}. \qquad (2.7.8)$$

In practice, it is often more convenient to read D directly from a graph of the cumulative periodogram than to use the above formula. Approximate critical values, D_c, for a test of white noise are tabulated below for a conventional set of significance levels. These values are taken from Stephens (1974), and are sufficiently accurate for all practical purposes.

	Level of significance		
	10%	5%	1%
$D_c(\sqrt{m'} + 0.12 + 0.11/\sqrt{m'})$	1.224	1.358	1.628

Example 2.11. Cumulative periodogram of an LH series
Figure 2.18 shows the cumulative periodogram of the LH series previously analysed in Examples 2.8a, 2.9 and 2.10a. Note that the 5% critical value of D is represented as a pair of parallel lines to the left and to the right of the line $y = x$. we therefore reject the white noise hypothesis at the 5% level of significance.

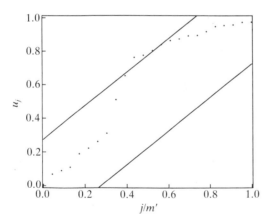

Fig. 2.18. The cumulative periodogram of the first of the three LH time series. The two straight lines correspond to the 5% critical value of the statistic D.

2.8 The connection between the correlogram and the periodogram

Although the correlogram and the periodogram appear to have quite different rationales, the following argument shows a close connection between them. For any Fourier frequency ω, we can write

$$I(\omega) = \left[\left\{ \sum_{t=1}^{n} y_t \cos(\omega t) \right\}^2 + \left\{ \sum_{t=1}^{n} y_t \sin(\omega t) \right\}^2 \right] \Big/ n$$

$$= \left[\left\{ \sum_{t=1}^{n} (y_t - \bar{y}) \cos(\omega t) \right\}^2 + \left\{ \sum_{t=1}^{n} (y_t - \bar{y}) \sin(\omega t) \right\}^2 \right] \Big/ n,$$

since $\sum_{t=1}^{n} \cos(\omega t) = \sum_{t=1}^{n} \sin(\omega t) = 0$, by Theorem 2.1. Expanding each squared term gives

$$nI(\omega) = \sum (y_t - \bar{y})^2 \{ \cos^2(\omega t) + \sin^2(\omega t) \}$$

$$+ \sum_{s \neq t} \sum (y_t - \bar{y})(y_s - \bar{y}) \{ \cos(\omega t) \cos(\omega s) + \sin(\omega t) \sin(\omega s) \}$$

$$= \sum (y_t - \bar{y})^2 + 2 \sum_{k=1}^{n-1} \sum_{t=k+1}^{n} (y_t - \bar{y})(y_{t-k} - \bar{y}) \cos(k\omega),$$

using standard trigonometric identities. Now, substituting the sample autocovariance coefficients defined at eqn. (2.5.1) into the above expression, we obtain

$$I(\omega) = g_0 + 2 \sum_{k=1}^{n-1} g_k \cos(k\omega), \tag{2.8.1}$$

which expresses the periodogram as the discrete Fourier transform of the sample autocovariance function. Finally, dividing throughout by g_0 defines the *normalized periodogram*,

$$I(\omega)/g_0 = 1 + 2 \sum_{k=1}^{n-1} r_k \cos(k\omega),$$

as the discrete Fourier transform of the correlogram.

   This result gives a hint as to why the periodogram has a wider value than as a device to detect strictly cyclic patterns in data and, incidentally, explains our choice of the constant of proportionality in eqn. (2.7.7). The normalized periodogram describes the overall pattern of serial dependence in a time series, in a way which is mathematically equivalent to the correlogram. However, this mathematical equivalence (by which we mean simply that the normalized periodogram is a known, invertible transformation of the correlogram) does not translate to statistical equivalence (by which we mean that statistical procedures based on the correlogram and on the periodogram typically use the underlying data in quite different ways and thereby highlight different aspects thereof). To illustrate the point, we conclude this section with a simple comparison between the cumulative periodogram and portmanteau tests for departure from white noise.

Table 2.1. Power estimates for tests of white noise against an
autoregressive alternative, $Y_t = \alpha Y_{t-1} + Z_t$

α	D	Q_1	Q_2	Q_5	Q_{10}
(a) 5% level of significance					
0.0	0.044	0.048	0.031	0.057	0.057
0.1	0.149	0.141	0.111	0.094	0.101
0.2	0.452	0.470	0.351	0.274	0.229
0.3	0.782	0.818	0.725	0.599	0.483
0.4	0.957	0.970	0.932	0.870	0.800
0.5	0.997	0.999	0.997	0.988	0.956
(b) 1% level of significance					
0.0	0.005	0.007	0.005	0.011	0.014
0.1	0.053	0.045	0.030	0.022	0.032
0.2	0.229	0.243	0.167	0.125	0.103
0.3	0.575	0.631	0.509	0.373	0.301
0.4	0.859	0.895	0.843	0.736	0.625
0.5	0.979	0.992	0.981	0.946	0.887

The comparison involves calculating the test statistics for simulated realizations of the first-order autoregressive process, $Y_t = \alpha Y_{t-1} + Z_t$, with Normally distributed Z_t. We simulated 1000 series of length $n = 100$ for each of $\alpha = 0.0,\ 0.1,\ 0.2,\ 0.3,\ 0.4,\ 0.5$. Table 2.1 gives the proportions of occasions out of 1000 on which white noise was rejected at the 5% and 1% critical values, using as test statistics:

(a) the cumulative periodogram ordinate statistic D (cf. eqn. (2.7.8)),
(b) the portmanteau statistic Q_m for $m = 1, 2, 5, 10$ (cf. eqn. (2.5.4)).

The results for $\alpha = 0.0$ are compatible with the prescribed significance levels: note that results for the various tests statistics are derived from the same set of simulations, and are not independent. For positive α, it is known from theoretical considerations that Q_1 is the most powerful of the five tests against this particular alternative to white noise. The results are again compatible with this, but suggest that D is competitive. On the other hand, the power of Q_m declines progressively as m increases. This highlights a practical difficulty with the portmanteau statistic, namely that the rather arbitrary choice of m affects its power characteristics. Choosing m small implicitly restricts the range of alternatives under consideration, whereas choosing m large risks a loss of power because the higher-order autocorrelations often contribute essentially no useful information.

3
Theory of stationary random processes

3.1 Notation and definitions

We begin this chapter by recalling some notation and definitions from Chapters 1 and 2.

A *random function* is a set of random variables $\{Y(t)\}$, in which the time-index, t can assume any real value. The *trend* of $\{Y(t)\}$ is the non-random function $\mu(t) = E[Y(t)]$. The *autocovariance function* of $\{Y(t)\}$ is

$$\gamma(t, s) = \text{cov}\{Y(t), Y(s)\} = E[\{Y(t) - \mu(t)\}\{Y(s) - \mu(s)\}].$$

The trend and autocovariance function together constitute the *second-order properties* of $\{Y(t)\}$. Second-order properties are the most common basis for statistical analysis of time-series data, and provide a complete description if the joint distribution of any finite collection of random variables $\{Y(t_1), \ldots, Y(t_n)\}$ is multivariate Normal.

The random function $\{Y(t)\}$ is *stationary* if $\mu(t) = \mu$ and $\gamma(t, s) = \gamma(|t - s|)$, i.e. the trend is constant and the covariance between the random variables $Y(t)$ and $Y(s)$ depends only on their separation in time. If $\{Y(t)\}$ is stationary, we define the *autocorrelation function*, $\rho(k) = \gamma(k)/\gamma(0)$.

A *random sequence* is a collection of random variables $\{Y_t\}$, in which the time-index assumes integer values only. The definitions given above for random functions still apply. However, in describing the properties of random sequences we shall usually adopt a subscript notation to emphasize the discrete nature of the time-index, for example μ_t rather than $\mu(t)$ and γ_k rather than $\gamma(k)$.

A *white noise sequence* is a particular type of random sequence. It consists of mutually independent random variables Y_t, each with mean zero and common variance, σ^2 say.

We use the single term *random process* to embrace random functions and sequences. The theory of random processes is both deep and extensive. In this book, we discuss this theory only to the extent that we require it for the later chapters on statistical methodology. An excellent starting point for the reader who would like a more detailed and mathematically rigorous account is the book by Priestley (1981).

Most of our discussion will be in terms of discrete-time models, i.e. random sequences rather than random functions. One very good, if pragmatic, reason for this is that most time-series data are presented as sequences of observations equally spaced in time, and can therefore be analysed conveniently using discrete-time models. However, it is important to recognize this practice for the convenience it undoubtedly is: Section 3.5 illustrates some of the complications which can arise when discrete-time processes are used as models for underlying continuous-time phenomena.

3.2 The spectrum of a stationary random process

Consider a stationary random sequence $\{Y_t\}$ with autocovariance function $\gamma_k = \text{cov}\{Y_t, Y_{t-k}\}$. The corresponding *autocovariance generating function* is the function

$$G(z) = \sum_{k=-\infty}^{\infty} \gamma_k z^k, \qquad (3.2.1)$$

whose argument, z, is a complex variable. If in eqn. (3.2.1) we now choose $z = e^{-i\omega}$, where ω is a real variable, we obtain the *spectrum* of $\{Y_t\}$,

$$f(\omega) = G(e^{-i\omega}) = \sum_{k=-\infty}^{\infty} \gamma_k e^{-ik\omega}. \qquad (3.2.2)$$

Because $\gamma_k = \gamma_{-k}$ and $e^{i\omega} + e^{-i\omega} = 2\cos(\omega)$ we can also write eqn. (3.2.2) as

$$f(\omega) = \gamma_0 + 2 \sum_{k=1}^{\infty} \gamma_k \cos(k\omega), \qquad (3.2.3)$$

revealing that the spectrum is a real-valued function.

If σ^2 denotes the variance of Y_t we can similarly define a *normalized spectrum*,

$$f^*(\omega) = f(\omega)/\sigma^2$$

$$= 1 + 2 \sum_{k=1}^{\infty} \rho_k \cos(k\omega).$$

Note that the normalized spectrum bears the same relationship to the autocorrelation function as does the spectrum to the autocovariance function. Note also that the normalized spectrum of a white noise sequence assumes a constant value 1 for all ω. This constant spectrum is analogous to the flat optical spectrum of white light, and is the source of the term 'white noise'.

The complex forms (3.2.1) and (3.2.2) are often convenient for theoretical analysis of stationary random sequences, but eqn. (3.2.3) is more illuminating in that it establishes the intimate connection between the spectrum of $\{Y_t\}$ and the periodogram of an observed time series $\{y_t\}$. In Section 2.8 we showed that the periodogram of $\{y_t\}$ can be written as

$$I(\omega) = g_0 + 2 \sum_{k=1}^{n-1} g_k \cos(k\omega),$$

where g_k is an estimate of γ_k. Comparing this with eqn. (3.2.3) we see that the periodogram is simply the empirical analogue of the spectrum, in which theoretical autocovariances γ_k are replaced by sample auto-covariances g_k, and the summation of covariance terms is necessarily truncated at one less than the length of the observed series.

Note from the form of eqn. (3.2.3) that $f(\omega) = f(-\omega)$ and $f(\omega) = f(\omega + 2\pi m)$ for all integers m. These features together imply that we need only define $f(\omega)$ for ω in the range 0 to π, a result which also has its empirical counterpart in the restriction of the periodogram to frequencies ω in the range 0 to π.

Definition (3.2.2) expresses the spectrum as the discrete Fourier transform of the autocovariance function. The converse relationship is that

$$\gamma_k = (2\pi)^{-1} \int_{-\pi}^{\pi} e^{ik\omega} f(\omega) \, d\omega \tag{3.2.4}$$

$$= \pi^{-1} \int_{0}^{\pi} \cos(k\omega) f(\omega) \, d\omega, \tag{3.2.5}$$

remembering that $f(\omega) = f(-\omega)$. In fact, eqn. (3.2.4) or (3.2.5) provides a characterization of the class of legitimate autocovariance functions. This characterization, known as Wold's Theorem essentially states that any non-negative valued function $f(\omega)$ on $(0, \pi)$ defines a legitimate spectrum, and γ_k is therefore a legitimate autocovariance function if and only if it can be expressed in the form (3.2.4) or (3.2.5) for some such $f(\omega)$. The proof is beyond the scope of this book but may be found, for example, in Priestley (1981, Chapter 4). The important point to note here is that the permissible form of γ_k is quite severely constrained, whereas that of $f(\omega)$ is not.

From a mathematical point of view, the spectrum and autocovariance function contain equivalent information concerning the underlying stationary random sequence $\{Y_t\}$. However, the spectrum also has a more tangible interpretation in terms of the inherent tendency for realizations of $\{Y_t\}$ to exhibit cyclic variations about the mean. Recall in particular that in Section 2.7 our initial motivation for the periodogram was not in terms of a mathematical transformation of the sample autocovariance function, but rather as a partitioning of the total sum of squares of the data into components due to harmonic regressions at various frequencies. An example helps to clarify this point.

Example 3.1. A first-order autoregressive process
Suppose that $\{Y_t\}$ is defined by

$$Y_t = \alpha Y_{t-1} + Z_t,$$

where $\{Z_t\}$ is a white noise sequence and $-1 < \alpha < 1$. We have already seen in Section 2.4 that the autocorrelation function of $\{Y_t\}$ is

$$\rho_k = \alpha^k \quad : \quad k = 0, 1, \ldots,$$

with $\rho_k = \rho_{-k}$. Thus, the normalized spectrum of $\{Y_t\}$ is

$$f^*(\omega) = \sum_{k=-\infty}^{\infty} \rho_k e^{-ik\omega}$$

$$= 1 + \sum_{k=1}^{\infty} \alpha^k e^{-ik\omega} + \sum_{k=1}^{\infty} \alpha^k e^{ik\omega}$$

$$= 1 + \frac{\alpha e^{-i\omega}}{1 - \alpha e^{-i\omega}} + \frac{\alpha e^{i\omega}}{1 - \alpha e^{i\omega}}$$

$$= \frac{(1 - \alpha e^{-i\omega})(1 - \alpha e^{i\omega}) + \alpha e^{-i\omega}(1 - \alpha e^{i\omega}) + \alpha e^{i\omega}(1 - \alpha e^{-i\omega})}{(1 - \alpha e^{-i\omega})(1 - \alpha e^{i\omega})}.$$

To simplify this expression we multiply out the various terms, noting that $e^{-i\omega} e^{i\omega} = 1$ and $e^{-i\omega} + e^{i\omega} = 2 \cos(\omega)$. This gives the final result,

$$f^*(\omega) = (1 - \alpha^2)\{1 - 2\alpha \cos(\omega) + \alpha^2\}^{-1}. \tag{3.2.6}$$

Figure 3.1 shows the normalized spectrum (3.2.6) for each of $\alpha = -0.5$, 0.5, and 0.9, together with a sample realization of the corresponding $\{Y_t\}$. For negative α, $f^*(\omega)$ is an increasing function of ω. We refer to this as a 'concentration of power at high frequencies', and the corresponding sample realization shows rapid oscillations, with a tendency for positive and negative values to alternate. For positive α, $f^*(\omega)$ is a decreasing function of ω, i.e. power is concentrated at low frequencies, and the sample realization reflects this in its tendency to take relatively long excursions above or below the zero level. These same simulated sequences were shown in Fig. 2.9 along with their corresponding autocorrelation functions.

Note, finally, that if σ^2 denotes the variance of the white noise sequence $\{Z_t\}$, and v the variance of $\{Y_t\}$, then it follows from the

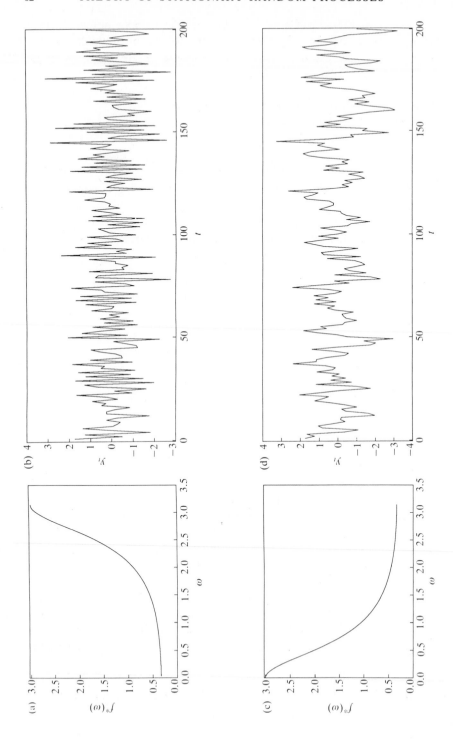

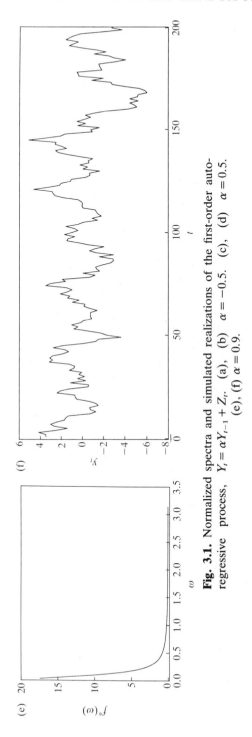

Fig. 3.1. Normalized spectra and simulated realizations of the first-order auto-regressive process, $Y_t = \alpha Y_{t-1} + Z_t$. (a), (b) $\alpha = -0.5$. (c), (d) $\alpha = 0.5$. (e), (f) $\alpha = 0.9$.

definition of $\{Y_t\}$ that

$$v = \alpha^2 v + \sigma^2,$$

or

$$v = \sigma^2/(1 - \alpha^2).$$

Thus, the (non-normalized) spectrum of $\{Y_t\}$ is

$$f(\omega) = \sigma^2 \{1 - 2\alpha \cos(\omega) + \alpha^2\}^{-1}.$$

The above results have obvious counterparts for stationary random functions $\{Y(t)\}$, the essential difference being that $f(\omega)$ is no longer periodic. We define the spectrum of $\{Y(t)\}$ as

$$f(\omega) = \int_{-\infty}^{\infty} \gamma(u)e^{-iu\omega}\, du.$$

This again has the property that $f(\omega) = f(-\omega)$, but it is no longer true that $f(\omega) = f(\omega + 2\pi m)$ for integer m. The physical explanation for this is that when a series is observed in continuous time, it is possible to identify separately effects at arbitrarily high frequencies, whereas when a series is observed at unit time intervals, effects at frequencies higher than $\omega = \pi$ are indistinguishable from lower-frequency effects. This phenomenon, known as 'aliasing', is illustrated by the following simple example.

Example 3.2. Aliasing
Consider a non-random function,

$$y(t) = \cos(t\omega),$$

where $\omega = \omega_0 + \pi$ and $0 < \omega_0 < \pi$. Suppose that this function is observed only at integer time points t. Since $\cos(x) = \cos(x + 2\pi t)$ for all integer t, and $\cos(x) = \cos(-x)$, the resulting observed sequence $\{y_t\}$ can be written as

$$\begin{aligned}
y_t &= \cos(t\omega) \\
&= \cos(t\omega - 2\pi t) \\
&= \cos(t\omega_0 - \pi t) \\
&= \cos(\pi t - t\omega_0) \\
&= \cos(t\omega'),
\end{aligned}$$

where $\omega' = \pi - \omega_0$ lies in the range 0 to π. Thus, frequency effects at $\omega > \pi$ and $\omega' < \pi$ are indistinguishable.

All of the above discussion begs the question of whether the infinite summation in eqn. (3.2.3) converges. The following example shows that it does not always do so.

Example 3.3. A divergent spectral series
Define a random sequence $\{Y_t\}$ by

$$Y_t = A \cos(\theta t) + B \sin(\theta t) + Z_t,$$

where $\{Z_t\}$ is a white noise sequence with variance σ^2, and A and B are independent random variables each with mean zero and variance τ^2. Now,

$$E[Y_t] = E[A] \cos(\theta t) + E[B] \sin(\theta t) + E[Z_t] = 0,$$

and

$$\begin{aligned} \text{var}(Y_t) &= E[A^2] \cos^2(t\theta) + E[B^2] \sin^2(t\theta) + E[Z_t^2] \\ &= \tau^2 \{\cos^2(t\theta) + \sin^2(t\theta)\} + \sigma^2 \\ &= \tau^2 + \sigma^2. \end{aligned}$$

Note that neither the mean nor the variance of Y_t involves t. Finally, for $s \neq t$,

$$\begin{aligned} \text{cov}\{Y_t, Y_s\} &= E[\{A \cos(\theta t) + B \sin(\theta t) + Z_t\}\{A \cos(\theta s) + B \sin(\theta s) + Z_s\}] \\ &= E[A^2] \cos(\theta t) \cos(\theta s) + E[B^2] \sin(\theta t) \sin(\theta s) \\ &= \tau^2 \cos\{\theta(t - s)\}, \end{aligned}$$

all other terms vanishing because they involve the expectations of products of independent zero-mean random variables.

The above results establish that $\{Y_t\}$ is a *stationary* random sequence with autocovariance function

$$\gamma_k = \begin{cases} \sigma^2 + \tau^2 & : \quad k = 0 \\ \tau^2 \cos(k\theta) & : \quad k \neq 0. \end{cases}$$

If the spectrum of $\{Y_t\}$ exists, then according to (3.2.3) it can be evaluated as

$$f(\omega) = \sigma^2 + \tau^2 + 2\tau^2 \sum_{k=1}^{\infty} \cos(k\theta) \cos(k\omega)$$

$$= \sigma^2 + \tau^2 + 2\tau^2 \sum_{k=1}^{\infty} [\cos\{k(\theta + \omega)\} + \cos\{k(\theta - \omega)\}].$$

Clearly, if $\theta = \omega$, then $\cos\{k(\theta - \omega)\} = 1$ for all k, and the summation is infinite. Thus, the spectrum of $\{Y_t\}$ can only be said to exist if we are prepared to adopt a special convention to allow $f(\omega)$ to be infinitely large at isolated values of ω.

The reader may reasonably take the view that Example 3.3 is no more than a mathematical contrivance. Were such a process ever to arise in practice, a sensible strategy would be to treat the *realized values* of A and

B as parameters to be estimated. This would amount to treating $\{Y_t\}$ as a non-stationary process with a simple harmonic trend, and the question of a spectrum for $\{Y_t\}$ would not then arise. Only if repeated realizations of $\{Y_t\}$ were observed would it be possible to make sensible inferences about the variance of A or B.

Example 3.3 is an instance of a *non-ergodic* stationary random process. Roughly speaking, a non-ergodic process is one whose parameters cannot all be estimated precisely from a single realization, however long; in this case, τ^2 cannot be so estimated. If we exclude non-ergodic processes from consideration, the summation in eqn. (3.2.2) or, equivalently, eqn. (3.2.3), always yields a finite value of $f(\omega)$, i.e. the spectrum is well defined. For a more rigorous discussion, see Priesley (1981, Section 4.9).

A final comment on the spectrum is a warning to the reader concerning a minor point of convention. Some authors define the spectrum to be a constant multiple of our $f(\omega)$; a common choice is $f(\omega)/(2\pi)$.

3.3 Linear filters

A *filter* is simply a transformation of one random sequence, $\{U_t\}$ say, into another, $\{Y_t\}$. A *linear filter* is a transformation of the form

$$Y_t = \sum_{j=-\infty}^{\infty} a_j U_{t-j}. \qquad (3.3.1)$$

Thus, each element of $\{Y_t\}$ is a linear combination of, possibly infinitely many, elements of $\{U_t\}$. If $\{U_t\}$ is stationary and only a finite number of the a_j are non-zero, then $\{Y_t\}$ is also stationary. When infinitely many a_j are non-zero, $\{Y_t\}$ may or may not be stationary, depending on the precise form of the a_j.

When $\{U_t\}$ and $\{Y_t\}$ are both stationary, the relationship between their spectra turns out to be strikingly simple. To derive this relationship, we note first that the autocovariance function of $\{Y_t\}$ is

$$\gamma_y(k) = \text{cov}\{Y_t, Y_{t-k}\}$$

$$= \text{cov}\left\{\sum_{j=-\infty}^{\infty} a_j U_{t-j} \sum_{l=-\infty}^{\infty} a_l U_{t-k-l}\right\}$$

$$= \sum_{j=-\infty}^{\infty} \sum_{l=-\infty}^{\infty} a_j a_l \gamma_u(k+l-j), \qquad (3.3.2)$$

where $\gamma_u(\cdot)$ denotes the autocovariance function of $\{U_t\}$. We now convert eqn. (3.3.2) to a relationship involving the autocovariance generating functions of $\{U_t\}$ and $\{Y_t\}$. Multiplying both sides of eqn.

(3.3.2) by z^k, rearranging terms and summing over k gives

$$\gamma_y(k)z^k = \sum_{j=-\infty}^{\infty} \sum_{l=-\infty}^{\infty} a_j z^j a_l z^{-l} \gamma_u(k+l-j)z^{k+l-j}$$

and

$$\sum_{k=-\infty}^{\infty} \gamma_y(k)z^k = \sum_{j=-\infty}^{\infty} a_j z^j \sum_{l=-\infty}^{\infty} a_l z^{-l} \sum_{k=-\infty}^{\infty} \gamma_u(k+l-j)u^{k+l-j}. \quad (3.3.3)$$

Clearly, the left-hand side of eqn. (3.3.3) is just $G_y(z)$, the auto-covariance generating function of $\{Y_t\}$. For the terms on the right-hand side, we define a function $A(z) = \sum_{j=-\infty}^{\infty} a_j z^j$ and change the final summation index to give the relationship

$$G_y(z) = A(z)A(z^{-1})G_u(z).$$

Finally, setting $z = \mathrm{e}^{-i\omega}$ gives the relationship between the spectra of $\{U_t\}$ and $\{Y_t\}$ as

$$f_y(\omega) = |a(\omega)|^2 f_u(\omega), \quad (3.3.4)$$

where

$$a(\omega) = \sum_{j=-\infty}^{\infty} a_j \mathrm{e}^{-ij\omega}$$

is called the *transfer function* of the linear filter. Note that eqn. (3.3.4) expresses the second-order properties of $\{Y_t\}$ as a simple product of two functions, one relating to the linear filter and the other to $\{U_t\}$, whereas for the analogous expression (3.3.2) in terms of covariance functions, the effects of the filter and of $\{U_t\}$ are combined in a rather complicated fashion.

Example 3.4. A three-point moving average of a first-order autoregressive process
Suppose that $\{U_t\}$ is defined by

$$U_t = \alpha U_{t-1} + Z_t,$$

where $\{Z_t\}$ is a white noise sequence with variance σ^2. In Example 3.1 we showed that the spectrum of $\{U_t\}$ is

$$f_u(\omega) = \sigma^2\{1 - 2\alpha \cos(\omega) + \alpha^2\}^{-1}.$$

Now, define a random sequence $\{Y_t\}$ by

$$Y_t = \tfrac{1}{3}(U_{t-1} + U_t + U_{t+1}),$$

i.e. a linear filter with

$$a_j = \begin{cases} \tfrac{1}{3} & : \quad j = -1, 0, 1 \\ 0 & : \quad \text{otherwise.} \end{cases}$$

The transfer function of this linear filter is

$$a(\omega) = (e^{i\omega} + 1 + e^{-i\omega})/3 = \{1 + 2\cos(\omega)\}/3$$

and its squared modulus is

$$|a(\omega)|^2 = \{1 + 4\cos(\omega) + 4\cos^2(\omega)\}/9$$
$$= \{3 + 4\cos(\omega) + 2\cos(2\omega)\}/9.$$

Thus, the spectrum of $\{Y_t\}$ is

$$f_y(\omega) = (\sigma^2/9)\{3 + 4\cos(\omega) + 2\cos(2\omega)\}\{1 - 2\alpha\cos(\omega) + \alpha^2\}^{-1}.$$

Figure 3.2 shows the functions $f_u(\omega)$, $|a(\omega)|^2$ and $f_y(\omega)$ for the case when $\sigma^2 = 1$ and $\alpha = -0.5$. Note in particular the comparison between $f_u(\omega)$ and $f_y(\omega)$. The effect of the moving average is both to reduce the total area under the spectrum, corresponding to a reduction in variance (a phenomenon well known in connection with means of independent random variables), and to alter the shape of the spectrum; the greater

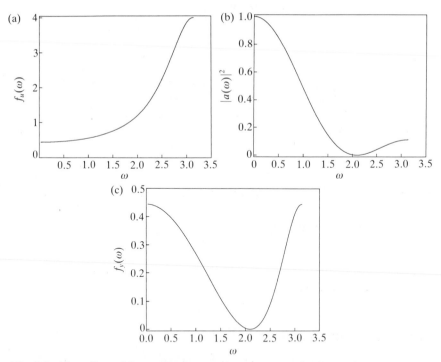

Fig. 3.2. The effect of linear filtering on the spectrum of a first-order autoregressive process, $U_t = -0.5U_{t-1} + Z_t$. (a) The spectrum of $\{U_t\}$. (b) The squared modulus of the transfer function. (c) The spectrum of the filtered process $\{Y_t\}$, where $Y_t = (U_{t-1} + U_t + U_{t+1})/3$.

concentration of power in the low-frequency range reflects the smoothing effect of the moving average. Note the connection between this discussion and that of Section 2.6 concerning the use of moving averages for trend removal.

An apparently special case of eqn. (3.3.1) is a random sequence $\{Y_t\}$ of the form

$$Y_t = \sum_{j=0}^{\infty} a_j Z_{t-j}, \tag{3.3.5}$$

where $\{Z_t\}$ is a white noise sequence with variance σ^2. This class of random sequences $\{Y_t\}$ is called the *general linear process* and, as its name implies, is less special than it at first appears. To see why, we shall derive its spectrum.

Firstly, the spectrum of $\{Z_t\}$ is

$$f_z(\omega) = \sigma^2 \quad : \quad 0 \leq \omega \leq \pi,$$

as is immediately clear from eqn. (3.2.3). It then follows from eqn. (3.3.4) that the spectrum of $\{Y_t\}$ is

$$f(\omega) = \sigma^2 |a(\omega)|^2, \tag{3.3.6}$$

where

$$a(\omega) = \sum_{j=0}^{\infty} a_j e^{-ij\omega}. \tag{3.3.7}$$

Combining eqns (3.3.6) and (3.3.7) we can write $f(\omega)$ as

$$f(\omega) = \sigma^2 \sum_{j=0}^{\infty} a_j e^{-ij\omega} \sum_{k=0}^{\infty} a_k e^{ik\omega}$$

$$= \sigma^2 \left[\sum_{j=0}^{\infty} a_j^2 + \sum_{j \neq k} \sum a_j a_k \exp\{i(k-j)\omega\} \right].$$

Now, using the fact that $e^{i\theta} + e^{-i\theta} = 2\cos(\theta)$, we can write

$$\sum_{j \neq k} \sum a_j a_k \exp\{i(k-j)\omega\} = 2 \sum_{j<k} a_j a_k \cos\{(k-j)\omega\}$$

and it follows that

$$f(\omega) = \sigma^2 \left\{ \sum_{j=0}^{\infty} a_j^2 + \sum_{m=1}^{\infty} \sum_{k=m}^{\infty} a_{k-m} a_k \cos(m\omega) \right\}$$

$$= b_0 + \sum_{m=1}^{\infty} b_m \cos(m\omega). \tag{3.3.8}$$

A standard result from classical Fourier analysis (Jeffreys and Jeffreys, 1956, Chapter 14) is that any real-valued, continuous function $f(\omega)$ which is both even and has period 2π (i.e. $f(\omega)=f(-\omega)$ and $f(\omega)=f(\omega+2\pi)$), is expressible in the form (3.3.8). It follows that, with regard to second-order properties, *any* stationary random sequence with a continuous spectrum has a representation as a general linear process (see Priestley, 1981, Section 4.9 for further discussion).

In practice, a general linear process is a useful model only when its coefficients a_j are expressible in terms of a finite number of parameters, which we can then hope to estimate from a set of data. A very rich class of models which satisfy this requirement is the autoregressive moving average process, which we shall discuss in some detail in the next section.

We conclude this section by establishing a simple condition on the coefficients a_j which ensures that the general linear process is stationary. This condition is implicit in the manipulations leading to eqn. (3.3.8). Essentially, we now repeat those manipulations but in the guise of an evaluation of the autocovariance function of $\{Y_t\}$. Note, firstly, that the definition (3.3.5) immediately implies that $E[Y_t]=0$, for all t. Hence, the covariance between Y_t and Y_s is

$$E[Y_tY_s] = E\left[\sum_{j=0}^{\infty} a_jZ_{t-j} \sum_{l=0}^{\infty} a_lZ_{s-l}\right]$$

$$= \sum_{j=0}^{\infty} \sum_{l=0}^{\infty} a_ja_lE[Z_{t-j}Z_{s-l}]. \tag{3.3.9}$$

Because $\{Z_t\}$ is white noise, the expectation term on the right-hand side of eqn. (3.3.9) is zero unless $t-j=s-l$, from which we deduce that, for $s \geq t$,

$$E[Y_tY_s] = \sigma^2 \sum_{j=0}^{\infty} a_ja_{j+s-t}. \tag{3.3.10}$$

Clearly, the right-hand side of eqn. (3.3.10) depends on t and s only through their difference, and $\{Y_t\}$ will therefore be stationary provided the infinite series $\sum_{j=0}^{\infty} a_ja_{j+k}$ converges for all non-negative integers k. Setting $k=0$, we certainly require that

$$\sum_{j=0}^{\infty} a_j^2 < \infty, \tag{3.3.11}$$

which is equivalent to the statement that $\{Y_t\}$ has finite variance. To see that the condition (3.3.11) is also sufficient, we use the fact that the correlation between any two random variables must be between -1 and $+1$. Thus,

$$|\text{cov}(Y_t, Y_{t-k})/\sqrt{\{\text{var}(Y_t)\,\text{var}(Y_{t-k})\}}| \leq 1.$$

But we already have that $\gamma_0 = \mathrm{var}(Y_t) = \mathrm{var}(Y_{t-k}) < \infty$, hence

$$|\gamma_\kappa| \leq \gamma_0 < \infty$$

and $\{Y_t\}$ is stationary, with autocovariance function

$$\gamma_k = \sigma^2 \sum_{j=0}^{\infty} a_j a_{j+k} \quad : \quad k = 0, 1, \ldots . \qquad (3.3.12)$$

3.4 The autoregressive moving average process

The basic building block for the autoregressive moving average process is a white noise sequence $\{Z_t\}$. A *moving average* process is simply a finite linear filter applied to $\{Z_t\}$, of the form

$$Y_t = Z_t + \sum_{j=1}^{q} \beta_j Z_{t-j}. \qquad (3.4.1)$$

In an *autoregressive* process, each Y_t is defined in terms of its predecessors Y_s, for $s < t$, by the equation

$$Y_t = \sum_{l=1}^{p} \alpha_l Y_{t-l} + Z_t. \qquad (3.4.2)$$

Note that eqn. (3.4.2) expresses $\{Z_t\}$ as the result of a finite linear filter applied to $\{Y_t\}$. In this sense, eqns (3.4.1) and (3.4.2) are identical, but with the roles of $\{Y_t\}$ and $\{Z_t\}$ reversed; this duality will recur. Finally, an *autoregressive moving average* process combines eqns (3.4.1) and (3.4.2) into a defining equation

$$Y_t = \sum_{l=1}^{p} \alpha_l Y_{t-l} + Z_t + \sum_{j=1}^{q} \beta_j Z_{t-j}. \qquad (3.4.3)$$

An obvious and convenient shorthand notation for this type of process is to write $Y_t \sim \mathrm{ARMA}(p, q)$. The integers p and q are called the *orders* of the process. A further abbreviation, if either of p or q is zero, is to write $Y_t \sim \mathrm{MA}(q)$ or $Y_t \sim \mathrm{AR}(p)$ as appropriate.

A slightly more general definition of an ARMA process incorporates a non-zero mean value, μ, say. The definining equation is then

$$Y_t - \mu = \sum_{l=1}^{p} \alpha_l (Y_{t-l} - \mu) + Z_t + \sum_{j=1}^{q} \beta_j Z_{t-j}. \qquad (3.4.4)$$

For the theoretical investigations in this chapter, nothing is lost and some economy of expression is gained if we assume $\mu = 0$, i.e. use eqn. (3.4.3) rather than eqn. (3.4.4).

3.4.1 *The backward shift operator: a convenient notational shorthand*

Because ARMA processes involve linear combinations of the elements of random sequences $\{Y_t\}$ and $\{Z_t\}$, it is useful to have a shorthand notation for the operation of shifting the time index. For this, we use the *backward shift operator B*, whereby BY_t is to be interpreted as Y_{t-1}. More usefully, powers of B are to be interpreted as successive applications of B; thus, for example, $B^2Y_t = B(BY_t) = BY_{t-1} = Y_{t-2}$. Finally, 1 indicates a null operation, $1Y_t = Y_t$, and mathematical functions of B are to be interpreted as their formal series expansions. For example,

$$(1 - B/2)^{-1}Y_t = \left\{\sum_{l=0}^{\infty}(B/2)^l\right\}Y_t = \sum_{l=0}^{\infty}2^{-l}Y_{t-l}.$$

Using this operator notation, the definition (3.4.3) for a zero-mean ARMA(p, q) process can be expressed in the more compact form,

$$\phi(B)Y_t = \theta(B)Z_t, \tag{3.4.5}$$

where $\phi(\cdot)$ and $\theta(\cdot)$ are polynomials of degree p and q respectively, and $\phi(0) = \theta(0) = 1$.

In eqn. (3.4.5), the side condition that $\phi(0) = \theta(0) = 1$ eliminates redundancy in the specification of $\{Y_t\}$ which arises through arbitrary re-scaling. For example, suppose that in eqn. (3.4.5), $\phi(B) = 2 - B$ and $\theta(B) = 3 + 2B$. Writing out the corresponding ARMA(1, 1) process explicitly gives

$$2Y_t - Y_{t-1} = 3Z_t + 2Z_{t-1}, \tag{3.4.6}$$

where $\{Z_t\}$ is a white-noise sequence with variance σ^2, say. Clearly, eqn. (3.4.6) can be written equivalently as

$$Y_t - \tfrac{1}{2}Y_{t-1} = Z_t' + \tfrac{2}{3}Z_{t-1}', \tag{3.4.7}$$

where now $\{Z_t'\}$ is a white-noise sequence with variance $(3\sigma/2)^2$. Reverting to the operator notation, eqn. (3.4.7) corresponds to the general form eqn. (3.4.5) with $\phi(B) = 1 - \tfrac{1}{2}B$, $\theta(B) = 1 + 2B/3$ and $\phi(0) = \theta(0) = 1$, as required.

3.4.2 *Second-order properties of the moving average process:* MA(q)

As already noted, the MA(q) process (3.4.1) is a form of linear filter applied to white noise, and its spectral properties can be deduced almost immediately. We first write eqn. (3.4.1) in the operator notation,

$$Y_t = \theta(B)Z_t,$$

where $\theta(\cdot)$ is a polynomial of degree q with $\theta(0) = 1$,

$$\theta(B) = 1 + \sum_{j=1}^{q} \beta_j B^j.$$

The spectrum of $\{Y_t\}$ can therefore be deduced immediately from eqn. (3.3.6) as

$$f(\omega) = \sigma^2 |\theta(e^{-i\omega})|^2$$

$$= \sigma^2 \left[\left\{ 1 + \sum_{j=1}^{q} \beta_j \cos(j\omega) \right\}^2 + \left\{ \sum_{j=1}^{q} \beta_j \sin(j\omega) \right\}^2 \right]. \qquad (3.4.8)$$

The autocovariance function can similarly be deduced from eqn. (3.3.8), making the appropriate substitutions between the general linear filter coefficients a_j and the moving average parameters β_j. However, a direct derivation is straightforward and may be illuminating.

Firstly, it is clear from eqn. (3.4.1) that $E[Y_t] = 0$ for all t. Thus, the autocovariance function is

$$\gamma_k = E[Y_t Y_{t-k}]$$

$$= E \left[\sum_{j=0}^{q} \beta_j Z_{t-j} \sum_{i=0}^{q} \beta_i Z_{t-k-i} \right],$$

with $\beta_0 = 1$. Using the linearity of the expectation operator we can write the above expression as

$$\gamma_k = \sum_{j=0}^{q} \sum_{i=0}^{q} \beta_j \beta_i E[Z_{t-j} Z_{t-k-i}].$$

But the expectation term is zero unless $t - j = t - k - i$, because $\{Z_t\}$ is a white noise sequence. Thus the only non-zero terms in the summation are of the form $\sigma^2 \beta_{i+k} \beta_i$ and we arrive at the result,

$$\gamma_k = \begin{cases} \sigma^2 \sum_{i=0}^{q-k} \beta_{i+k}\beta_i & : \quad k = 0, 1, \ldots, q \\ 0 & : \quad k > q. \end{cases} \qquad (3.4.9)$$

The 'cut-off' in the autocovariances after lag $k = q$ is a characteristic property of the MA(q) process. Note that the non-zero autocorrelations for the MA(q) process take the form

$$\rho_k = \left(\sum_{i=0}^{q-k} \beta_{i+k}\beta_i \right) \bigg/ \left(\sum_{i=0}^{q} \beta_i^2 \right). \qquad (3.4.10)$$

Example 3.5. $Y_t \sim \mathrm{MA}(1)$

The case $q = 1$ gives the simplest illustration of the above results. Writing $\beta_1 = \beta$, the spectrum of $\{Y_t\}$ is

$$f(\omega) = \sigma^2[\{1 + \beta \cos(\omega)\}^2 + \{\beta \sin(\omega)\}^2]$$
$$= \sigma^2[1 + 2\beta \cos(\omega) + \beta^2].$$

The variance of $\{Y_t\}$ is $\sigma^2(1 + \beta^2)$. The only non-zero autocorrelation is

$$\rho_1 = \beta/(1 + \beta^2).$$

Note that ρ_1 takes the same sign as β and is constrained to be in the interval $-\frac{1}{2}$ to $\frac{1}{2}$. Figure 3.3 shows the spectrum of $f(\omega)$ for $\sigma^2 = 1$ and each of $\beta = -1.0$ and 1.0, together with a sample realization of each process. When $\beta > 0$, the large values of $f(\omega)$ occur for small values of ω, corresponding to the smoothing effect of the moving average operation on the white noise sequence, whereas when $\beta < 0$ the large values of $f(\omega)$ near $\omega = \pi$ reflect a tendency for successive Y_t to alternate in sign. These remarks parallel the commentary to Example 3.1 in which qualitatively similar behaviour was noted for the AR(1) process.

3.4.3 *Second-order properties of the autoregressive process:* AR(p)

The AR(p) process (3.4.2) expresses the current Y_t as a linear combination of past Y_s and a white noise term, Z_t. Whilst this is an intuitively natural description, and the source of the term 'autoregressive', its second-order properties are most easily derived by noting that it formally expresses the white noise sequence $\{Z_t\}$ as a linear filter of $\{Y_t\}$.

Having made the above observation, and recalling that the spectrum of white noise is a constant, σ^2, we immediately deduce from eqn. (3.3.4) that the spectrum of $\{Y_t\}$, $f(\omega)$, say, satisfies

$$|\phi(e^{-i\omega})|^2 f(\omega) = \sigma^2,$$

where $\phi(\cdot)$ is the polynomial

$$\phi(B) = 1 - \sum_{l=1}^{p} \alpha_l B^l.$$

Explicitly, this gives

$$f(\omega) = \sigma^2 \left[\left\{ 1 - \sum_{i=1}^{p} \alpha_l \cos(l\omega) \right\}^2 + \left\{ \sum_{l=1}^{p} \alpha_l \sin(l\omega) \right\}^2 \right]^{-1}. \quad (3.4.11)$$

The above derivation begs the non-trivial question of whether the AR(p) process is in fact stationary. The difficulty arises because the term in square brackets in eqn. (3.4.1) may be zero for some ω and

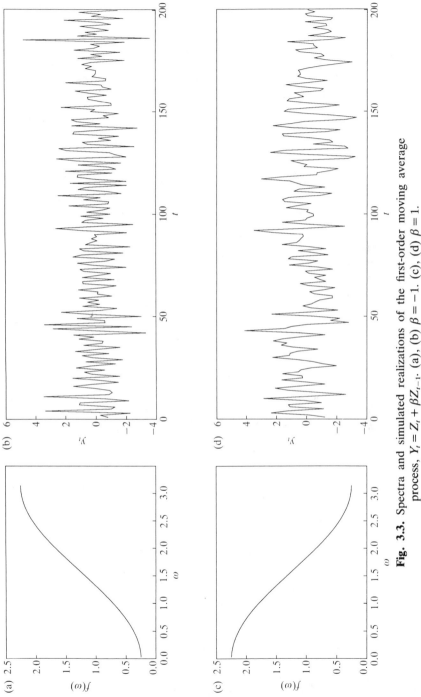

Fig. 3.3. Spectra and simulated realizations of the first-order moving average process, $Y_t = Z_t + \beta Z_{t-1}$: (a), (b) $\beta = -1$. (c), (d) $\beta = 1$.

particular values of the parameters σ_l. For example, if $p = 1$ and $\alpha_1 = 1$ eqn. (3.4.11) gives

$$f(\omega) = \tfrac{1}{2}\sigma^2[1 - \cos(\omega)]^{-1},$$

which is infinite at $\omega = 0$.

To establish the precise conditions under which the AR(p) process is stationary, we shall use its representation as a general linear process,

$$Y_t = \{\phi(B)\}^{-1}Z_t$$
$$= \left(\sum_{j=0}^{\infty} a_j B^j\right)Z_t = \sum_{j=0}^{\infty} a_j Z_{t-j},$$

for suitable coefficients a_j. Then, according to the results of Section 3.3, $\{Y_t\}$ is stationary if and only if $\sum_{j=0}^{\infty} a_j^2 < \infty$. The following theorem translates this condition into an explicit condition on the polynomial $\phi(u) = 1 - \sum_{l=1}^{p} \alpha_l u^l$.

Theorem 3.1. *The* AR(p) *process* $\{Y_t\}$ *defined by* $\phi(B)Y_t = Z_t$, *where* $\{Z_t\}$ *is white noise, is stationary if and only if all roots of the equation* $\phi(u) = 0$ *have modulus greater than one.*

Proof. We write the polynomial $\phi(u)$ explicitly as

$$\phi(u) = 1 - \sum_{l=1}^{p} \alpha_l u^l = \prod_{l=1}^{p} (1 - b_l u), \tag{3.4.12}$$

so that the roots of the equation $\phi(u) = 0$ are the (possibly complex) numbers b_l^{-1}. To avoid a technical complication in the proof, we assume that no two of the b_l are equal. It follows that we can choose constants c_l (which depend on the b_l) such that

$$\{\phi(u)\}^{-1} = \sum_{l=1}^{p} c_l/(1 - b_l u).$$

For u sufficiently small, so that $|b_l u| < 1$, a power series expansion of each term in the above expression gives

$$\{\phi(u)\}^{-1} = \sum_{l=0}^{p} c_l \sum_{j=0}^{\infty} b_l^j u^j = \sum_{j=0}^{\infty} a_j u^j, \tag{3.4.13}$$

where

$$a_j = \sum_{l=1}^{p} c_l b_l^j.$$

The last expression in eqn. (3.4.13) gives the representation of $\{Y_t\}$ as a general linear process,

$$Y_t = \sum_{j=0}^{\infty} a_j Z_{t-j}.$$

Thus, $\{Y_t\}$ is stationary if and only if $\sum_{j=0}^{\infty} a_j^2 < \infty$.

Now, suppose that all the coefficients b_l in eqn. (3.4.12) are less than one in modulus, i.e. $|b_l| \le b < 1$ for all l. Then,

$$a_j^2 \le \left(\sum_{l=1}^p c_l \right)^2 b^{2j}$$

and

$$\sum_{j=0}^\infty a_j^2 \le \left(\sum_{l=1}^p c_l \right)^2 \bigg/ (1 - b^2) < \infty.$$

Conversely, if max $|b_l| \ge 1$, the term in a_j corresponding to that particular b_l eventually dominates as j increases, and $\sum_{j=0}^\infty a_j^2$ diverges. Thus, the necessary and sufficient condition for stationarity of $\{Y_t\}$ is that all the coefficients b_l should be strictly less than one in absolute value. Since the roots of the polynomial equation $\phi(u) = 0$ are b_l^{-1}, this completes the proof of the theorem.   ∎

The autocorrelation function of a stationary $AR(p)$ process can be derived by inversion of the spectrum. However, as with the $MA(q)$ process, a direct derivation is straightforward, and in this case is of some independent interest since it suggests a method of parameter estimation for autoregressive processes.

Multiplying both sides of eqn. (3.4.2) by Y_{t-k} gives

$$Y_t Y_{t-k} = \sum_{l=1}^p \alpha_l Y_{t-l} Y_{t-k} + Z_t Y_{t-k}.$$

For $k > 0$, Z_t and Y_{t-k} are independent. Taking expectations on both sides of the above equation therefore gives

$$\gamma_k = \sum_{l=1}^p \alpha_l \gamma_{k-l} \quad : \quad k = 1, 2, \ldots . \tag{3.4.14}$$

Dividing both sides of eqn. (3.4.14) by γ_0 then gives an analogous set of equations in terms of the autocorrelation coefficients,

$$\rho_k = \sum_{l=1}^p \alpha_l \rho_{k-l} \quad : \quad k = 1, 2, \ldots , \tag{3.4.15}$$

known as the Yule–Walker equations. Our present interest in these equations is as a means of calculating the ρ_k from the α_l. However, they can also be used to infer values of α_l corresponding to an observed set of sample autocorrelation coefficients. This inferential use of the Yule–Walker equations will be discussed in Chapter 6.

Example 3.6. $Y_t \sim AR(2)$
The AR(2) process,

$$Y_t = \alpha_1 Y_{t-1} + \alpha_2 Y_{t-2} + Z_t,$$

can generate a surprisingly rich variety of second-order properties, depending on the numerical values of the parameters α_1 and α_2. We shall use the Yule–Walker equations to obtain an explicit expression for the autocorrelation function.

The Yule–Walker equations express the ρ_k as the solution of a system of linear difference equations. We look for solutions of the form $\rho_k = \lambda^k$. Substitution of this expression into eqn. (3.4.15) gives

$$\lambda^k = \alpha_1 \lambda^{k-1} + \alpha_2 \lambda^{k-2},$$

or

$$\lambda^2 - \alpha_1 \lambda - \alpha_2 = 0. \tag{3.4.16}$$

This quadratic equation has two roots,

$$\lambda = \{\alpha_1 \pm \sqrt{(\alpha_1^2 + 4\alpha_2)}\}/2, \tag{3.4.17}$$

and the nature of the solution depends on whether these results are real and distinct ($\alpha_1^2 + 4\alpha_2 > 0$), real and coincident ($\alpha_1^2 + 4\alpha_2 = 0$) or complex ($\alpha_1^2 + 4\alpha_2 < 0$).

Case 1. $\alpha_1^2 + 4\alpha_2 > 0$.
If λ_1 and λ_2 are the real numbers given by eqn. (3.4.17), then any expression of the form $\rho_k = a\lambda_1^k + b\lambda_2^k$ will satisfy the Yule–Walker equations (3.4.15) for $k \geq 2$. To determine the unique solution for all k we note firstly that $\rho_0 = 1$, whence $b = 1 - a$, and secondly that $\rho_1 = \alpha_1 + \alpha_2 \rho_1$, whence $\rho_1 = \alpha_1/(1 - \alpha_2) = a\lambda_1 + (1 - a)\lambda_2$.

As a specific illustration, we take $\alpha_1 = 0.2$ and $\alpha_2 = 0.1$. Then, $\alpha_1^2 + 4\alpha_2 = 0.44$, $\lambda_1 = 0.3816$, $\lambda_2 = -0.2316$, $\rho_1 = 0.2222$ and

$$a = (0.2222 + 0.2316)/(0.3816 + 0.2316) = 0.5999.$$

Note that both λ_1 and λ_2 are less than one in absolute value, as required for stationarity of $\{Y_t\}$. The autocorrelation function follows as

$$\rho_k = 0.5999(0.3816)^k + 0.4001(-0.2316)^k.$$

Figure 3.4 shows this autocorrelation function and the corresponding spectrum, taking $\sigma^2 = 1$ for convenience.

Case 2. $\alpha_1^2 + 4\alpha_2 = 0$.
The quadratic equation (3.4.16) now has a pair of coincident roots, $\lambda = \alpha_1/2$. The solution to the Yule–Walker equations takes the form $\rho_k = (a + bk)\lambda^k$. The initial conditions $\rho_0 = 1$ and $\rho_1 = \alpha_1/(1 - \alpha_2)$ now imply that $a = 1$ and $(1 + b)\alpha_1/2 = \alpha_1/(1 - \alpha_2)$.

As a specific illustration, we take $\alpha_1 = 1$, $\alpha_2 = -0.25$. Then, $\lambda = 0.5$,

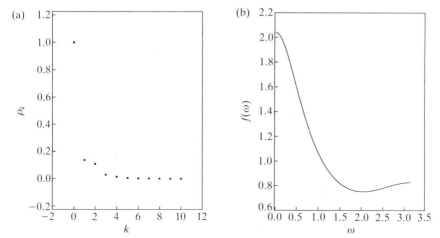

Fig. 3.4. The autocorrelation function and spectrum of the AR(2) process, $Y_t = 0.2Y_{t-1} + 0.1Y_{t-2} + Z_t$. (a) Autocorrelation function. (b) Spectrum.

$b = 0.6$ and

$$\rho_k = (1 + 0.6k)0.5^k.$$

Figure 3.5 shows this autocorrelation function and the corresponding spectrum, again taking $\sigma^2 = 1$. Note in particular the increased concentration of power into the low-frequency range by comparison with Fig. 3.4.

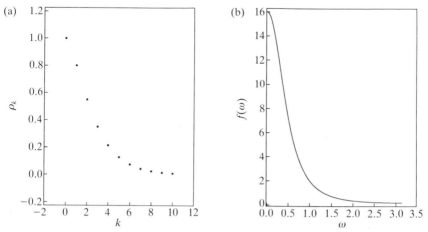

Fig. 3.5. The autocorrelation function and spectrum of the AR(2) process, $Y_t = Y_{t-1} - 0.25Y_{t-2} + Z_t$. (a) Autocorrelation function. (b) Spectrum.

Case 3. $\alpha_1^2 + 4\alpha_2 < 0$.

The roots of the quadratic equation (3.4.16) are now a complex conjugate pair, $\lambda_1 = re^{i\theta}$ and $\lambda_2 = re^{-i\theta}$, where $r = -\alpha_2 > 0$ and $\theta = \tan^{-1}\{(-\alpha_1^2 - 4\alpha_2)/\alpha_1\}$, to be interpreted as lying in the range $\pi/2$ to π if α_1 is negative. The autocorrelation function is

$$\rho_k = r^k(ae^{ik\theta} + be^{-ik\theta}),$$

where a and b are complex numbers determined by the initial conditions as before. Note that r must be less than one for stationarity of $\{Y_t\}$. Also, the solution for ρ_k must be real-valued, and takes the form

$$\rho_k = r^k\{A\cos(k\theta) + B\sin(k\theta)\}.$$

As a specific illustration, we take $\alpha_1 = 1.0$, $\alpha_2 = -0.5$. Then, $r = 0.5$ and $\theta = \pi/4$. The initial condition $\rho_0 = 1$ gives $A = 1$; $\rho_1 = 0.6667$ gives

$$0.667 = 0.5\{\cos(\pi/4) + B\sin(\pi/4)\},$$

i.e. $B = -0.5286$ and

$$\rho_k = 0.5^k\{\cos(\pi k/4) - 0.5286\sin(\pi k/4)\}.$$

Figure 3.6 shows this autocorrelation function and the corresponding spectrum. The peak in the spectrum points to quasi-cyclic behaviour of the underlying AR-process.

To conclude this example, Fig. 3.7 shows a sample realization of each

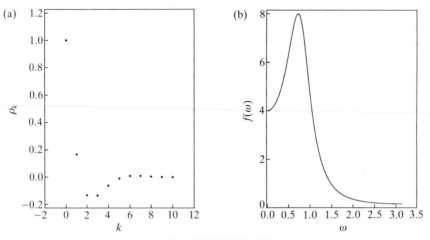

Fig. 3.6. The autocorrelation function and spectrum of the AR(2) process, $Y_t = Y_{t-1} - 0.5Y_{t-2} + Z_t$. (a) Autocorrelation function. (b) Spectrum.

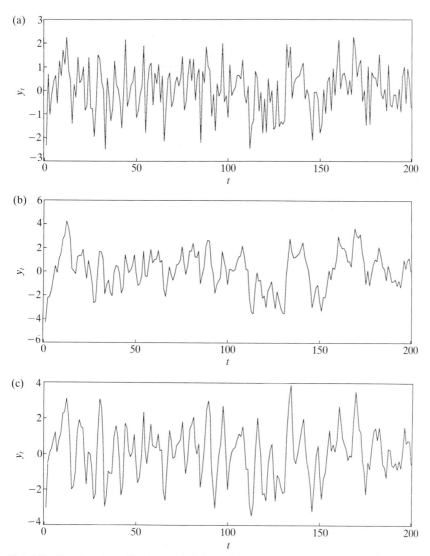

Fig. 3.7. Simulated realizations of the AR(2) process, $Y_t = \alpha_1 Y_{t-1} + \alpha_2 Y_{t-2} + Z_t$.
(a) $\alpha_1 = 0.2$, $\alpha_2 = 0.1$. (b) $\alpha_1 = 1$, $\alpha_2 = -0.25$. (c) $\alpha_1 = 1$, $\alpha_2 = -0.5$.

of the three specific AR(2) processes considered. It is instructive to look first at the three realizations and then at the corresponding autocorrelation functions and spectra, to gain some understanding of how the distinctive appearances of the three sample realizations are reflected in the second-order properties of the corresponding processes.

3.4.4 *Second-order properties of the autoregressive moving average process:* ARMA(p, q)

Simple adaptations of the techniques in the previous two subsections can be used to derive the second-order properties of the ARMA-process $\{Y_t\}$ defined equivalently by eqn. (3.4.3) or eqn. (3.4.5), that is

$$Y_t = \sum_{l=1}^{p} \alpha_l Y_{t-l} + Z_t + \sum_{j=1}^{q} \beta_j Z_{t-j} \tag{3.4.3}$$

or

$$\phi(B)Y_t = \theta(B)Z_t. \tag{3.4.5}$$

Using results on the linear filter, we can immediately deduce from eqn. (3.4.5) that the spectrum, $f(\omega)$, must satisfy

$$|\phi(e^{-i\omega})|^2 f(\omega) = \sigma^2 |\theta(e^{-i\omega})|^2.$$

Thus,

$$f(\omega) = \sigma^2 |\theta(e^{-i\omega})|^2 |\phi(e^{-i\omega})|^{-2}$$

$$= \sigma^2 \left[\left\{ 1 + \sum_{j=1}^{q} \beta_j \cos(j\omega) \right\}^2 + \left\{ \sum_{j=1}^{q} \beta_j \sin(j\omega) \right\}^2 \right]$$

$$\times \left[\left\{ 1 - \sum_{l=1}^{p} \alpha_l \cos(l\omega) \right\}^2 + \left\{ \sum_{l=1}^{p} \alpha_l \sin(l\omega) \right\}^2 \right]^{-1},$$

assuming that $\{Y_t\}$ is stationary. The form of $f(\omega)$ suggests that the values of the α_l are critical in determining stationarity, and this is indeed the case. The precise condition is the same as for the autoregressive progress $\phi(B)Y_t = Z_t$, namely that all the roots of $\phi(u) = 0$ must be greater than one in absolute value.

The easiest way to obtain the autocorrelation function in specific cases is to express $\{Y_t\}$ as a general linear process,

$$Y_t = \{\phi(B)\}^{-1}\theta(B)Z_t = \left(\sum_{j=0}^{\infty} a_j B^j \right) Z_t,$$

the coefficients a_j being determined by a formal power series expansion of $\{\phi(B)\}^{-1}$.

Example 3.7. $Y_t \sim$ ARMA$(1, 1)$
The process $\{Y_t\}$ defined by

$$Y_t = \alpha Y_{t-1} + Z_t + \beta Z_{t-1}$$

is stationary if $-1 < \alpha < 1$. In this case, we write

$$Y_t = (1 - \alpha B)^{-1}(1 + \beta B)Z_t$$

$$= \sum_{j=0}^{\infty} \alpha^j B^j (1 + \beta B)Z_t$$

$$= Z_t + \sum_{j=1}^{\infty} (\alpha + \beta)\alpha^{j-1} Z_{t-j}.$$

Now, according to eqn. (3.3.12), the variance of $\{Y_t\}$ is

$$\gamma_0 = \sigma^2 \left\{ 1 + \sum_{j=1}^{\infty} (\alpha + \beta)^2 \alpha^{2(j-1)} \right\}$$

$$= \sigma^2 \{ 1 + (\alpha + \beta)^2 / (1 - \alpha^2) \}.$$

Also, for $k \geqslant 1$,

$$\gamma_k = \sigma^2 \left\{ (\alpha + \beta)\alpha^{k-1} + \sum_{j=1}^{\infty} (\alpha + \beta)\alpha^{k+j-1}(\alpha + \beta)\alpha^{j-1} \right\}$$

$$= \sigma^2 \left\{ (\alpha + \beta)\alpha^{k-1} + (\alpha + \beta)^2 \alpha^k / (1 - \alpha^2) \right\}.$$

Note that for $k \geqslant 2$, $\gamma_k = \alpha\gamma_{k-1}$, and therefore that $\rho_k = \alpha\rho_{k-1}$. This is reminiscent of the exponentially decaying autocorrelation function of an AR(1) process. The distinction is that for the ARMA(1, 1) process, $\rho_1 \neq \alpha$. To emphasize this point, we illustrate the case $\alpha = 0.25$, $\beta = 1.0$. Figure 3.8 shows the autocorrelation function and spectrum of $\{Y_t\}$.

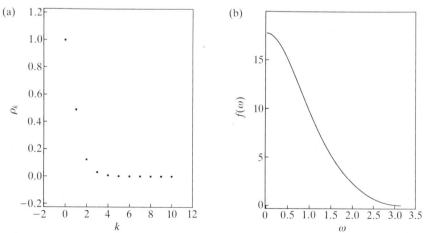

Fig. 3.8. The autocorrelation function and spectrum of the ARMA(1, 1) process, $Y_t = 0.25Y_{t-1} + Z_t + Z_{t-1}$. (a) Autocorrelation function. (b) Spectrum.

3.4.5 *Invertibility*

To introduce the notion of invertibility, consider the MA(1) process,

$$Y_t = Z_t + \beta Z_{t-1}.$$

The only non-zero autocorrelation associated with this process is

$$\rho_1 = \beta/(1 + \beta^2).$$

Suppose that we replace β by β^{-1}. Then,

$$\rho_1 = \beta^{-1}/(1 + \beta^{-2}) = \beta/(1 + \beta^2),$$

as before. To resolve this ambiguity, we define $\{Y_t\}$ in terms of the backward-shift operator,

$$Y_t = (1 + \beta B)Z_t,$$

and 'invert' this definition to give

$$Z_t = (1 + \beta B)^{-1}Y_t, \tag{3.4.18}$$

expressing $\{Z_t\}$ as a linear filter of $\{Y_t\}$, rather than *vice versa*. Does this inversion make physical sense? Taking a formal power series expansion of $(1 + \beta B)^{-1}$ in eqn. (3.4.18) gives

$$Z_t = \sum_{j=0}^{\infty} \beta^j Y_{t-j}.$$

Thus, if we think of each Z_t as being determined by present and past Y_s, rather than vice versa, we see that the remote past has a vanishingly small influence if and only if $-1 < \beta < 1$.

Quite generally, we say that the ARMA process $\{Y_t\}$ defined by

$$\phi(B)Y_t = \theta(B)Z_t$$

is *invertible* if the roots of the polynomial equation $\theta(u) = 0$ are all greater than one in absolute value. Note the exact duality with the condition on $\phi(u)$ for stationarity of $\{Y_t\}$.

3.4.6 *Why are ARMA processes useful?*

Given that AR, MA and ARMA processes can all be regarded as special cases of the general linear process, the reader might well ask why the three classes of processes are of independent interest. In rare instances, particular ARMA processes can be justified in terms of the basic data-generating mechanisms. More frequently, they are used in a purely empirical manner as a means of summarizing a time series by a few well-chosen summary statistics, namely the parameters of an ARMA process, with no implication that the resulting 'model' has any scientific basis.

It follows that ARMA processes are usually of limited value in a modelling exercise for which the primary purpose is to develop an understanding of the mechanisms which generate the data to hand. They are of considerably greater value when modelling is no more than a means to an end: for example, in forecasting or when the serial correlation structure has the status of a set of nuisance parameters. Their usefulness in this regard stems from the ability of the spectrum of an ARMA(p, q) process to assume a wide variety of shapes without requiring either p or q to be particularly large. An immediate consequence is that the second-order properties of a stationary time series can often be well approximated by an ARMA process which is economical in its use of parameters.

3.5 Sampling and accumulation of stationary random functions

Most time-series methodology has been developed with a view to analysing random sequences $\{Y_t\}$; the use of ARMA processes as a generally applicable class of empirical models is typical in this regard. However, it is often the case that a random sequence $\{Y_t\}$ is derived from an underlying random function $\{X(t)\}$ in continuous time, either by *sampling*, whereby $Y_t = X(t)$ for integer t only, or by *accumulation*, whereby $Y_t = \int_{t-1}^{t} X(s)\,ds$. These two operations have different implications for the autocovariance structure of $\{Y_t\}$, and this in turn has implications for the interpretation of discrete-time models, like ARMA processes, which make no reference to the underlying phenomenon in continuous time.

Let $\gamma_x(k)$ and $\gamma_y(k)$ denote the autocovariance functions of $\{X(t)\}$ and $\{Y_t\}$, respectively. If $\{Y_t\}$ is a sampled version of $\{X(t)\}$, i.e. $Y_t = X(t)$, then clearly $\gamma_y(k) = \gamma_x(k)$, and the only distinction between the two autocovariance functions is that $\gamma_x(k)$ is defined for all real k and $\gamma_y(k)$ for integer k only. However, if $\{Y_t\}$ is an accumulated version of $X(t)$, i.e. $Y_t = \int_{t-1}^{t} X(s)\,ds$, then

$$\gamma_y(k) = \text{cov}\left\{\int_{t-1}^{t} X(s)\,ds, \int_{t-k-1}^{t-k} X(u)\,du\right\}, \qquad (3.5.1)$$

the explicit form of which may differ materially from $\gamma_x(k)$.

Example 3.8. Interpretation of discrete-time models

Suppose that the random function $X(t)$ has autocovariance function $\gamma_x(k) = \sigma^2 \exp(-\lambda k) : k \geq 0$. If $\{Y_t\}$ is a sampled version of $\{X(t)\}$ its autocovariance function is $\gamma_y(k) = \sigma^2 \alpha^k : k = 0, 1, \ldots$, where $\alpha = e^{-\lambda}$. Since $\mathrm{var}(Y_t) = \sigma^2$, the autocorrelation function of $\{Y_t\}$ is $\rho_k = \alpha^k$, which is precisely the autocorrelation function of an AR(1) process.

If, instead, $\{Y_t\}$ is an accumulated version of $\{X(t)\}$, its autocovarance function is given by eqn. (3.5.1). Interchanging the order of the covarance operator and the integration in eqn. (3.5.1) gives

$$\gamma_y(k) = \int_{t-1}^{t} \int_{t-k-1}^{t-k} \mathrm{cov}\{X(s), X(u)\}\, du\, ds$$

$$= \int_{t-1}^{t} \int_{t-k-1}^{t-k} \gamma_x(|s-u|)\, du\, ds. \qquad (3.5.2)$$

Now, if $k \geq 1$, $s - u$ is necessarily non-negative and eqn. (3.5.2) becomes

$$\gamma_y(k) = \sigma^2 \int_{t-1}^{t} \int_{t-k-1}^{t-k} \exp\{-\lambda(s-u)\}\, du\, ds$$

$$= \sigma^2 \int_{t-1}^{t} e^{-\lambda s}\, ds \int_{t-k-1}^{t-k} e^{\lambda u}\, du.$$

Straightforward calculations then give the explicit result

$$\gamma_y(k) = (\sigma^2/\lambda^2) e^{\lambda} (1 - e^{-\lambda}) e^{-\lambda k} \quad : \quad k \geq 1.$$

Note in particular that $\gamma_y(k+1) = \alpha \gamma_y(k)$, where $\alpha = e^{-\lambda}$. To convert this into a statement about autocorrelation structure, we need to evaluate

$$\gamma(0) = \sigma^2 \int_{t-1}^{t} \int_{t-1}^{t} \exp(-\lambda |s-u|)\, du\, ds$$

$$= 2\sigma^2 \int_{t-1}^{t} \int_{t-1}^{s} \exp\{-\lambda(s-u)\}\, du\, ds$$

$$= (2\sigma^2/\lambda)\{1 - (1 - e^{-\lambda})/\lambda.$$

It follows that the autocorrelation function of $\{Y_t\}$ satisfies $\rho_{k+1} = \alpha \rho_k$, but with $\rho_1 \neq \alpha$, which is the autocorrelation function of an ARMA(1, 1) process.

Example 3.8 shows that apparently different discrete-time models can arise simply through the distinction between sampling and accumulation of the same phenomenon in continuous time. More generally, when discrete-time models are applied to sampled *or* accumulated data, their parameters can be awkward to interpret. This may become a major problem if we wish to compare results from series sampled at, or accumulated over, different time intervals.

This discussion is intended as a caution against the uncritical use of discrete-time models for continuous-time phenomena. It is not meant to deny the value of these models as empirical descriptors of observed data. On the contrary, Example 3.8 gives a specific example of an accumulated time series for which an ARMA(1, 1) process is an appropriate descriptive model.

3.6 Implications of autocorrelation for elementary statistical methods

Even when the autocorrelation structure of a time series is not of direct interest, it has an indirect bearing on the interpretation of any summary statistics calculated from the time series. In this section, we illustrate some of the problems which can arise if autocorrelation is not properly recognized.

Perhaps the most widely used of all elementary statistical methods is estimation of a mean value. If $Y_1, \ldots, Y_n$ are mutually independent random variables with common mean μ and variance σ^2, then

$$\bar{Y} = n^{-1} \sum_{i=1}^{n} Y_i$$

is the 'best linear unbiased' estimator for μ in the sense that $E[\bar{Y}] = \mu$ and $\text{var}(\bar{Y}) = \sigma^2/n$ is smaller than the variance of any other linear combination $L = \sum_{i=1}^{n} a_i Y_i$ for which $E[L] = \mu$. Also,

$$S^2 = (n-1)^{-1} \sum_{i=1}^{n} (Y_i - \bar{Y})^2$$

is an unbiased estimator for σ^2.

Let $\bar{y}$ and s^2 denote the sample mean and variance, i.e. the realized values of $\bar{Y}$ and S^2. The quantity $s/\sqrt{n}$ is called the *standard error* of $\bar{y}$. It is usual to quote $\bar{y} \pm 2s/\sqrt{n}$ as an approximate 95% confidence interval for μ. The strict justification for this rests on an additional assumption, namely that the Y_i are Normally distributed; for then, $(\bar{Y} - \mu)\sqrt{n}/S$ follows a t-distribution on $n-1$ degrees of freedom, and this in turn is well approximated by a standard Normal distribution, the quality of the approximation improving with n.

It is well known that the above procedure for constructing a confidence interval for μ is robust to non-Normality of the Y_i, essentially because S^2 is still an unbiased estimator for σ^2 and the central limit theorem guarantees that the sampling distribution of $\bar{Y}$ is approximately Normal. The consequences of autocorrelation amongst the Y_i are less reassuring.

If the Y_i constitute a stationary random sequence, it remains true that $\bar{Y}$ is an unbiased estimator for μ. However, its variance depends on the autocorrelation structure of $\{Y_t\}$. Specifically,

$$\text{var}(\bar{Y}) = n^{-2} \, \text{var}\left(\sum_{i=1}^{n} Y_i\right)$$

$$= n^{-2} \sum_{i=1}^{n} \sum_{j=1}^{n} \text{cov}(Y_i, Y_j)$$

$$= (\sigma^2/n)\left\{1 + 2n^{-1} \sum_{k=1}^{n-1} (n-k)\rho(k)\right\}, \qquad (3.6.1)$$

where $\rho(k)$ denotes the autocorrelation function of $\{Y_t\}$. Table 3.1 gives some numerical illustrations of the extent to which the autocorrelation structure can inflate the variance of $\bar{Y}$, using the autocorrelation function $\rho(k) = \exp(-\lambda k)$. For sufficiently small λ, corresponding to strong autocorrelation, the variance inflation can be substantial, and in any event increases with the length n of the time series. Note that if the observations are obtained by irregular sampling of a stationary random function, so that $Y_i = Y(t_i)$, then eqn. (3.6.1) must be modified to

$$\text{var}(\bar{Y}) = (\sigma^2/n)\left\{1 + 2n^{-1} \sum_{i=2}^{n} \sum_{j=1}^{i-1} \rho(|t_i - t_j|)\right\}. \qquad (3.6.2)$$

Although $\bar{Y}$ remains unbiased for μ, it is no longer the *best* linear unbiased estimator. For observations unequally spaced in time, it is intuitively clear that observations remote in time from their neighbours contain more information than observations closely spaced in time if, as is typical, $\rho(k)$ is a decreasing function of k.

A formal analysis proceeds as follows. Let **1** denote a vector with each

Table 3.1. Variance inflation of the sample mean due to autocorrelation: $\text{var}(\bar{Y})/(\sigma^2/n)$ for a time series of n observations at unit time-spacing and autocorrelation function $\rho(k) = \exp(-\lambda k)$

λ	n			
	5	10	25	100
0.1	4.29	7.38	12.68	18.02
0.2	3.73	5.72	8.05	9.54
0.5	2.64	3.30	3.77	4.00
1.0	1.80	1.98	2.09	2.15
2.0	1.24	1.28	1.30	1.31
5.0	1.01	1.01	1.01	1.01

Table 3.2. Efficiency of the sample mean relative to the best linear unbiased estimator of the population mean, based on observations $y(t_i): i = 1, \ldots, n$ from a stationary process with autocorrelation function $\rho(k) = \exp(-\lambda k)$

(a) Equally spaced observations: $t(i) = i$

		n	
λ	5	10	25
0.1	0.971	0.934	0.897
0.2	0.958	0.921	0.915
0.5	0.955	0.944	0.964
1.0	0.976	0.979	0.989
2.0	0.996	0.997	0.999
5.0	1.000	1.000	1.000

(b) Unequally spaced observations: $n = 5$, $t_1 = 1.0$, $t_2 = 5.0 - d$, $t_3 = 5.0$, $t_4 = 5.0 + d$, $t_5 = 9.0$

		d		
λ	2.0	1.0	0.5	0.1
0.1	0.958	0.930	0.915	0.902
0.2	0.952	0.906	0.878	0.852
0.5	0.976	0.909	0.854	0.798
1.0	0.996	0.951	0.882	0.790
2.0	1.000	0.991	0.945	0.813
5.0	1.000	1.000	0.997	0.878

of its elements equal to 1. Suppose that the vector $\mathbf{Y} = (Y_1, \ldots, Y_n)'$ has mean vector $\mu\mathbf{1} = (\mu, \ldots, \mu)'$ and variance matrix V, a symmetric n by n matrix with $(i, j$th) element $v_{ij} = \text{cov}(Y_i, Y_j)$. Then, the best linear unbiased estimator for μ is

$$\hat{\mu} = (\mathbf{1}'V^{-1}\mathbf{1})^{-1}\mathbf{1}'V^{-1}\mathbf{Y},$$

with variance

$$\text{var}(\hat{\mu}) = (\mathbf{1}'V^{-1}\mathbf{1})^{-1}.$$

Table 3.2 shows some values of the efficiency of $\bar{Y}$ relative to $\hat{\mu}$,

$$e = \text{var}(\hat{\mu})/\text{var}(\bar{Y}),$$

again assuming an autocorrelation function $\rho(k) = \exp(-\lambda k)$. Note that even when the Y_i are equally spaced in time, $\bar{Y}$ is slightly inefficient because $\hat{\mu}$ gives extra weight to the end-points in the sequence. However, the clear message from Table 3.2 is that in general, $\bar{Y}$ is only slightly inefficient.

A more awkward question concerns the standard error of $\bar{Y}$. An immediate problem is that S^2 is no longer an unbiased estimator for σ^2. The general expression for $E[S^2]$ is easy to derive, if somewhat unilluminating. We write

$$S^2 = (n-1)^{-1} \sum_{i=1}^{n} \left(Y_i - n^{-1} \sum_{j=1}^{n} Y_j \right)^2$$

$$= (n-1)^{-1} n^{-2} \sum_{i=1}^{n} \left(\sum_{j=1}^{n} a_{ij} Y_j \right)^2,$$

where

$$a_{ij} = \begin{cases} n-1 & : \quad j = i \\ -1 & : \quad j \neq i. \end{cases}$$

Then,

$$S^2 = (n-1)^{-1} n^{-2} \sum_{i=1}^{n} T_i^2,$$

where

$$T_i = \sum_{j=1}^{n} a_{ij} Y_j.$$

Now,

$$E[T_i] = \mu \sum_{j=1}^{n} a_{ij} = 0,$$

from which it follows that

$$E[T_i^2] = \operatorname{var}(T_i) = \sum_{j=1}^{n} \sum_{k=1}^{n} a_{ij} a_{ik} \operatorname{cov}(Y_j, Y_k)$$

and, finally, that

$$E[S^2] = (n-1)^{-1} n^{-2} \sum_{i=1}^{n} E[T_i^2]$$

$$= (n-1)^{-1} n^{-2} \sum_{i=1}^{n} \sum_{j=1}^{n} \sum_{k=1}^{n} a_{ij} a_{ik} \operatorname{cov}(Y_j, Y_k).$$

Table 3.3 shows some values of $E[S^2]/\sigma^2$, again using the autocorrelation function $\rho(k) = \exp(-\lambda k)$ and assuming observations equally spaced in time. Note how $E[S^2]$ is consistently smaller than σ^2. The explanation for this lies in the behaviour of sample realizations of positively

Table 3.3. $E[S^2]/\sigma^2$ for series of n observations at unit time-spacing and autocorrelation function $\rho(k) = \exp(-\lambda k)$

	n		
λ	5	10	25
0.1	0.177	0.291	0.513
0.2	0.317	0.475	0.706
0.5	0.589	0.744	0.885
1.0	0.800	0.891	0.955
2.0	0.940	0.969	0.988
5.0	0.997	0.999	1.000

autocorrelated random processes. In particular, in Example 3.1 we noted how a sample realization of a positively autocorrelated first-order autoregressive process tends to take relatively long excursions above or below its mean level of zero. One consequence of this is that the sample mean of a short sequence of realized values can be substantially different from the underlying population mean so that the sample variance S^2, which measures deviations about the sample mean, can be substantially smaller than the population variance, which measures deviations about the population mean.

Perhaps more to the point, Table 3.4 shows some values of the quantity

$$f = \sqrt{\mathrm{var}(\bar{Y})/(E[S^2]/n)\}}$$

which loosely speaking, represents the factor by which we can expect

Table 3.4. Values of $f = \sqrt{\{\mathrm{var}(\bar{Y})/(E[S^2]/n)\}}$ for series of n observations at unit time-spacing and autocorrelation function $\rho(k) = \exp(-\lambda k)$

	n		
λ	5	10	25
0.1	4.92	5.04	4.97
0.2	3.43	3.47	3.38
0.5	2.12	2.11	2.06
1.0	1.50	1.49	1.48
2.0	1.15	1.15	1.15
5.0	1.01	1.01	1.01

confidence intervals based on a false assumption of independence to be too narrow. These results show that if the autocorrelation is strong, confidence intervals based on the assumption of independence can be seriously misleading.

Analogous calculations can be performed for a general linear model,

$$\mathbf{Y} = X\boldsymbol{\theta} + \mathbf{U},$$

where $\mathbf{U} = (U_1, \ldots, U_n)'$ is generated by a zero-mean, stationary random process. The general conclusion is again that the ordinary least squares estimators

$$\tilde{\boldsymbol{\theta}} = (X'X)^{-1}X'\mathbf{Y}$$

are usually adequate for point estimation, whereas for interval estimation and, more generally, inference concerning $\boldsymbol{\theta}$, the autocorrelation structure of $\mathbf{U}$ must be taken into account. For a specific illustration, we have simulated 1000 samples of length n from a stationary AR(1) process with $\alpha = 0.5$, and for each sample computed the t-statistic to test for the significance of a linear time-trend, under the false assumption of independence. Figure 3.9 compares the empirical frequency distributions of these t-statistics with the probability density function of the t-distribution on $n - 2$ degrees of freedom, for each of $n = 10$ and $n = 25$. The empirical frequency distributions have been smoothed using a kernel estimate, with the smoothing parameter chosen optimally for an underlying Normal distribution (Silverman, 1986, Chapter 3). In each case, the empirical distribution is substantially more dispersed than the corresponding t-distribution, indicating that naïve application of a t-test would often give a spuriously significant result.

3.7   Further reading

We have already mentioned Priestley (1981) as an excellent general introduction to the theory of stationary random processes. Box and Jenkins (1970) give a very detailed discussion of ARMA processes and their second-order properties. Nicholls and Quinn (1982) discuss random coefficient autoregressive processes in which, as their name suggests, the parameters of an AR process are assumed to be generated stochastically. This approach is related to state space modelling, which in turn derives from the Kalman filter (Kalman, 1960; Meinhold and Singpurwalla, 1983). In the Kalman filter, the distribution of the observations y_t conditional on a set of parameters is governed by a general linear model, whilst the values of the parameters evolve in time according to a multivariate stochastic process. In recent years there has been increasing interest in the development of non-Gaussian models. One of the most

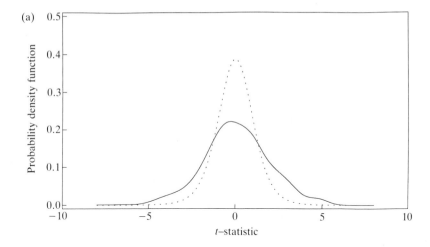

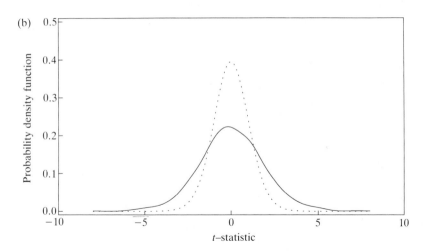

Fig. 3.9. The effect of autocorrelation on the sampling distribution of the *t*-statistic to test for a linear time-trend. —, Smoothed empirical distribution; - - - -, *t*-distribution on $n - 2$ degrees of freedom. (a) $n = 10$. (b) $n = 25$.

fruitful developments has been the extension of the state-space modelling approach to the non-Gaussian case. See, for example, West *et al.* (1985), Kitagawa (1987), Martin and Raftery (1987) and references therein. Examples of more specialized non-Gaussian modelling include Smith and West (1983) on renal transplant monitoring, Stern and Coe (1984) on the analysis of rainfall data, Lawrance and Lewis (1985) on the construction of autoregressive processes with exponential marginal distributions and Diggle and Zeger (1989) on a model for pulsatile time series motivated by data on hormonal concentrations in blood samples.

4
Spectral analysis

4.1 Introduction

In Section 2.7 we introduced the periodogram as a way of representing the variability in a time series in terms of harmonic components at various frequencies. We defined the periodogram ordinate at a particular frequency ω to be proportional to the squared amplitude of the corresponding cosine wave, $\alpha \cos(\omega t) + \beta \sin(\omega t)$, fitted to the data by least squares. This suggested that the periodogram could be used to identify important harmonic components in time-series data. In fact, as foreshadowed in Section 2.8, the periodogram has wider application. There, we pointed out that the periodogram is intimately connected to the sample autocovariance function, in that the one is an invertible mathematical transformation of the other. Put in statistical terms, the sample autocovariance function and the periodogram are alternative ways of summarizing the second-order properties of a time series. In Chapter 3 we explored this connection in a theoretical setting, showing that the autocovariance function and spectrum of a stationary random process are similarly related, the one being the Fourier transform of the other.

It follows from the above that a statistical analysis of the second-order properties of time-series data can be conducted in either of two ways, according to whether the correlogram or the periodogram is taken as the starting point. The former is called 'analysis in the time domain', and the latter 'analysis in the frequency domain' or, more simply, *spectral analysis*.

At first sight, it would seem natural to use the periodogram of a time series as an estimator for the spectrum of an underlying stationary process. It turns out that the periodogram itself is usually an unsatisfactory estimator, but that its deficiencies can be overcome by judicious smoothing of the sequence of periodogram ordinates. We treat this topic in Sections 4.2–4.7.

In Section 4.8 we consider the problem of comparing estimated spectra from two or more series. This gives a time-series analogue of a one-way analysis of variance, the data for analysis consisting of time series obtained under each of several different experimental conditions. Example 1.2, in which we might want to compare the pattern of fluctuations in LH during the early and late follicular phases of the menstrual cycle, provides a specific illustration.

In Section 4.9 we outline the use of spectral analysis in the fitting of parametric models to time-series data.

Finally, in Section 4.10 we discuss the strengths and weaknesses of spectral analysis, particularly in the context of analysing the relatively short series which prevail in biological applications.

One point worth emphasizing at the outset is that spectral analysis is only appropriate for *stationary* time-series data. In practice, this will mean either that the data are genuinely stationary, or that any trends in the data have been eliminated before the analysis, for example using the methods described in Section 2.2.

4.2 The periodogram revisited

In Sections 2.7 and 2.8 we first developed the periodogram from the point of view of a harmonic regression model, and later established its connection with the sample autocovariance function. We now take as our starting point the idea that an observed time series $\{y_t : t = 1, \ldots, n\}$ is a realization of a stationary random process $\{Y_t\}$ with spectrum $f(\omega)$ which we wish to estimate.

One possible estimator for $f(\omega)$ is the periodogram ordinate $I(\omega)$. Assuming ω to be a *Fourier frequency*, i.e. of the form $\omega_j = 2\pi j/n$ for a positive integer $j < n/2$, we note the following equivalent formulae for $I(\omega)$,

$$I(\omega) = n^{-1} \left[\left\{ \sum_{t=1}^{n} y_t \cos(\omega t) \right\}^2 + \left\{ \sum_{t=1}^{n} y_t \sin(\omega t) \right\}^2 \right], \tag{4.2.1}$$

$$I(\omega) = n^{-1} \left[\left\{ \sum_{t=1}^{n} (y_t - \bar{y}) \cos(\omega t) \right\}^2 + \left\{ \sum_{t=1}^{n} (y_t - \bar{y}) \sin(\omega t) \right\}^2 \right] \tag{4.2.2}$$

and

$$I(\omega) = g_0 + 2 \sum_{k=1}^{n-1} g_k \cos(k\omega) \tag{4.2.3}$$

where $\bar{y}$ is the sample mean and g_k the kth sample autocovariance of the series $\{y_t\}$. Equation (4.2.3) provides a reason for supposing that the periodogram might be used as an estimator for the spectrum, eqn. (4.2.1) is a more convenient computing formula, whilst the equivalence of eqns (4.2.1) and (4.2.2) is an immediate consequence of Theorem 2.1, part (i).

Our aims in the remainder of this section are to establish the sampling distribution of the periodogram ordinates $I(\omega)$, and to discuss the implications of this for estimation of the spectrum. Before stating our main result in general terms, it is instructive to explore the special case in which the underlying process $\{Y_t\}$ is Gaussian white noise, i.e. the random variables Y_t are mutually independent and Normally distributed, with common mean zero and variance σ^2.

We write eqn. (4.2.1) as

$$nI(\omega) = \{A(\omega)\}^2 + \{B(\omega)\}^2, \qquad (4.2.4)$$

where

$$A(\omega) = \sum_{t=1}^{n} Y_t \cos(\omega t)$$

and

$$B(\omega) = \sum_{t=1}^{n} Y_t \sin(\omega t).$$

Clearly, when $\{Y_t\}$ is Gaussian white noise, $A(\omega)$ and $B(\omega)$ are zero mean, Normally distributed random variables. Furthermore,

$$\operatorname{var}\{A(\omega)\} = \sigma^2 \sum_{t=1}^{n} \cos^2(\omega t) = (n\sigma^2)/2$$

and

$$\operatorname{var}\{B(\omega)\} = \sigma^2 \sum_{t=1}^{n} \sin^2(\omega t) = (n\sigma^2)/2$$

by Theorem 2.1, part (ii). Also,

$$\operatorname{cov}\{A(\omega), B(\omega)\} = E\left[\sum_{t=1}^{n} \sum_{s=1}^{n} Y_t Y_s \cos(\omega t) \sin(\omega s)\right]$$

$$= \sigma^2 \sum_{t=1}^{n} \cos(\omega t) \sin(\omega t) = 0,$$

by the Corollary to Theorem 2.1. It follows that $A(\omega)\sqrt{\{2/(n\sigma^2)\}}$ and $B(\omega)\sqrt{\{2/(n\sigma^2)\}}$ are independent standard Normal random variables, and therefore that $2[\{A(\omega)\}^2 + \{B(\omega)\}^2]/(n\sigma^2)$ is distributed as chi-squared on two degrees of freedom. Finally, substitution of this last result into eqn. (4.2.4) gives $I(\omega) \sim \sigma^2 \chi_2^2/2$ and, in particular,

$$E[I(\omega)] = \sigma^2$$
$$\operatorname{var}[I(\omega)] = \sigma^4.$$

Recall that the spectrum of white noise is $f(\omega) = \sigma^2$, for all ω. We have therefore established, at least in this special case, that the periodogram is an *unbiased* but *inconsistent* estimator for $f(\omega)$: unbiased because $E[I(\omega)] = f(\omega)$, inconsistent because $\operatorname{var}\{I(\omega)\}$ does not tend to zero as n tends to infinity. To complete the picture, we remark that for any two different Fourier frequencies, $\omega_j = 2\pi j/n$ and $\omega_k = 2\pi k/n$, $I(\omega_j)$ and $I(\omega_k)$ are statistically independent. The proof follows essentially the same argument as the proof that $A(\omega)$ and $B(\omega)$ are independent.

The inconsistency of $I(\omega)$ is at first sight surprising, but is readily explained by the fact that as n increases the data are being used to calculate proportionately more *independent* quantities $I(\omega)$, with the consequence that the precision attained for any one of them remains unchanged. Equivalently, the inconsistency of $I(\omega)$ as an estimator for $f(\omega)$ relates to its interpretation as a regression sum of squares on *two* degrees of freedom, irrespective of the value of n, within an orthogonal decomposition of the total sum of squares.

A similar result holds when $\{Y_t\}$ is any stationary Gaussian process, namely:

Theorem 4.1. *Let $\{Y_t\}$ be a stationary Gaussian process with spectrum $f(\omega)$, $\{y_t : t = 1, \ldots, n\}$ a partial realization of $\{Y_t\}$ and $I(\omega)$ the periodogram of $\{y_t\}$. Let $\omega_j = 2\pi j/n$ for positive integers $j < n/2$. Then in the limit $n \to \infty$,*

(a) $I(\omega_j) \sim f(\omega_j) \chi_2^2/2$
(b) $I(\omega_j)$ *and* $I(\omega_k)$ *are independent for all* $k \neq j$.

For a proof of Theorem 4.1 see Priestley (1981, Section 6.2). Note that if n is even, the theorem holds for the extreme Fourier frequency $\omega = \pi$ with the following modification to part (a):
(a) $I(\pi) \sim f(\pi) \chi_1^2$.
The reduction from two degrees of freedom to one is because $B(\pi)$ in eqn. (4.2.4) is identically zero.

Example 4.1. The periodogram of a simulated white noise series
In Fig. 2.17 we showed a simulated Gaussian white noise series of length $n = 200$, and its associated periodogram. In Fig. 4.1 we reproduce the periodogram, together with the periodogram of the first 50 observations only. Note firstly that the fluctuations in the periodogram ordinates are not systematically larger for the subseries of 50 observations than for the complete series: this is, of course, a direct consequence of the fact that the variance of each $I(\omega)$ does not depend on the length of the series from which it is calculated. Secondly, several of the periodogram ordinates are substantially greater than the value of 3.0 which is the 5% critical value for the sampling distribution of each $I(\omega)$, but these large periodogram ordinates are scattered haphazardly through the frequency range: this is a consequence of the independence of periodogram ordinates at different frequencies. The critical value 3.0 gives a legitimate test for the significance of a harmonic component of variation at single,

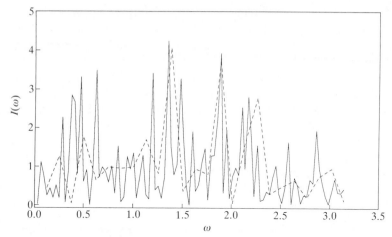

Fig. 4.1. The periodogram of a simulated Gaussian white noise series of length $n = 200$. ———, Periodogram of complete series: – – –, periodogram of first 50 observations.

prespecified frequency ω, but is clearly inappropriate for assessing general departure from white noise. However, the independence of the periodogram ordinates makes the construction of legitimate tests a relatively straightforward task, which we now consider.

4.3 Periodogram-based tests of white noise

In this section, we consider how the periodogram might be used to construct statistical tests for departure from white noise. Even when the white noise hypothesis is implausible *a priori*, it can provide a useful benchmark in the preliminary stages of data analysis, especially when the time series in question is relatively short: if the data are insufficient to allow rejection of white noise, there is probably little point in subjecting them to a more detailed analysis. Sometimes, the white noise hypothesis *is* of interest in its own right. In particular, many time-series models, including all ARMA processes, incorporate an embedded white noise sequence in their definition. It follows that tests of white noise can sometimes be used as approximate tests of the adequacy of such models. For example, if we want to test the adequacy of a first-order autoregressive process, $Y_t = \alpha Y_{t-1} + Z_t$, as a model for an observed series $\{y_t : t = 1, \ldots, n\}$, we can proceed as follows. Define a residual series $\{z_t : t = 2, \ldots, n\}$ by $z_t = y_t - \hat{\alpha} y_{t-1}$, where $\hat{\alpha}$ is an estimate of α. If the autoregressive model holds, the residual series $\{z_t\}$ is approximately a partial realization of white noise, the approximation arising only through the estimation of α. It follows that a test of white noise applied to $\{z_t\}$ is, in effect, a test of the adequacy of the autoregressive model.

In Section 2.7 we presented one periodogram-based test, the cumulative periodogram ordinate test. Here, we give the theoretical justification for that test, after first describing an older-established test based on the maximum periodogram ordinate. Both tests strictly assume that the data $\{y_t : t = 1, \ldots, n\}$ form a realization of *Gaussian* white noise. However, they remain valid to a good approximation for non-Gaussian data when n is large, because the central limit theorem then ensures that the quantities $A(\omega)$ and $B(\omega)$ in eqn. (4.2.4) are approximately Normally distributed, irrespective of the distribution of the y_t.

For both the maximum and cumulative periodogram ordinate tests, we consider the periodogram ordinates $I_j = I(2\pi j/n)$ for $j = 1, 2, \ldots, m$, where m is the largest integer strictly less than $n/2$. For n even, we exclude $I(\pi)$ because it has a different sampling distribution from that of all the other I_j. Then, Theorem 4.1 asserts that under the white noise hypothesis, the I_j form an independent random sample from a scaled χ_2^2-distribution, with distribution function

$$G(u) = P\{I_j \leq u\} = 1 - \exp(-u/\sigma^2).\qquad(4.3.1)$$

How should we test for departure from eqn. (4.3.1)? If the suspected source of departure is the presence of a single harmonic component at unknown frequency ω, a natural test statistic is the *maximum periodogram ordinate*,

$$T = \max_{j=1,\ldots,m} (I_j).$$

From eqn. (4.3.1) and the mutual independence of the I_j, we deduce that under the white noise hypothesis, the distribution function of T is

$$\begin{aligned}H(t) &= \{G(t)\}^m \\ &= \{1 - \exp(-t/\sigma^2)\}^m\end{aligned}\qquad(4.3.2)$$

In practice, σ^2 is almost invariably unknown. To obtain an approximate test, we can substitute the sample variance, $s^2 = \sum_{t=1}^{n} (y_t - \bar{y})^2/(n-1)$, for the unknown σ^2 in eqn. (4.3.2) in order to compute the attained significance level $1 - H(t)$. Fisher (1929) introduced the statistic T and derived the exact distribution function of $T_0 = T/\{\sum_{j=1}^{m} I_j/m\}$, the ratio of the maximum to average periodogram ordinate. In particular, he showed that

$$P\{T_0 > mx\} = \sum_{p=1}^{r} [m!/\{p!\,(m-r)!\}](-1)^{p-1}(1-px)^{m-1},\qquad(4.3.3)$$

where r is the largest integer less than x^{-1}.

Example 4.2. Simulated white noise

Using the same data as in Example 4.1, we computed the maximum periodogram ordinate, t, and the sample variance, s^2, for the full series of $n = 200$ observations and the subseries of length 50. We then compared the results of tests of white noise on the basis of

(a) the approximate test using eqn. (4.3.2) with σ^2 estimated by s^2,
(b) Fisher's exact test using eqn. (4.3.3).

For the subseries, $t = 4.0536$, $s^2 = 1.1139$ and the attained significance levels for the two tests were 0.4722, and 0.3197, respectively. For the full series, $t = 4.2305$, $s^2 = 0.9618$ and the corresponding attained significance levels were 0.7062 and 0.7975. Even for the full series, there is a noticeable discrepancy between the approximate and exact tests. The discrepancy could be important in cases of marginal significance, emphasizing the practical value of the exact test.

In the jargon of spectral analysis, white noise corresponds to a 'completely flat spectrum', $f(\omega) = \sigma^2$ for all ω, and the maximum periodogram ordinate test is designed to be sensitive against the alternative of a single 'sharp peak' at unknown frequency. If the actual form of departure from white noise involves a non-flat but smooth spectrum, it would seem preferable to use a test statistic which examines sequences of periodogram ordinates, since we would expect to find a preponderance of relatively large values in frequency regions where $f(\omega)$ is large, and *vice versa*. One such test is the cumulative periodogram ordinate test, as described in Section 2.7.

Recall that the test statistic is

$$D = \max_{j=1,\ldots,m'} \{\max(|U_j - j/m'|, |U_j - (j-1)/m'|)\}, \qquad (4.3.4)$$

where $m' = m - 1$, $U_j = C_j/C_m$ and $C_j = \sum_{k=1}^{j} I_k$. This is precisely the Kolmogorov–Smirnov statistic to test the hypothesis that the U_j are an ordered random sample from the uniform distribution on $(0, 1)$. A table of critical values of D was given at the end of Section 2.7.

The theoretical basis for the cumulative periodogram ordinate test lies in the connection between the exponential distribution (4.3.1) and the stochastic point process known as a Poisson process. A *Poisson process* generates a sequence of points P_i with the property that the intervals, $P_i - P_{i-1}$, between successive points are an independent random sample from an exponential distribution. It follows that the cumulative periodogram ordinates, $C_1, C_2, \ldots, C_m$, define a set of m successive points in a Poisson process, following a point of the process at the origin. Now, another property of the Poisson process is that, conditional on observing a point of the process at the origin and the mth subsequent point at C_m, the positions of the intervening $m' = (m - 1)$ points are an independent random sample from the uniform distribution on $(0, C_m)$. The C_j are the ordered values of these intervening m' points, and we deduce that the quantities $U_j = C_j/C_m : j = 1, \ldots , m'$ are an ordered random sample from the uniform distribution on $(0, 1)$.

It follows from the above that any goodness-of-fit statistic for the uniform distribution on $(0, 1)$ can be adapted to give a test of white noise. In particular, the Kolmogorov–Smirnov statistic is the maximum absolute difference between the theoretical distribution function of the uniform distribution on $(0, 1)$, $F(u) = u$, and the empirical distribution of the U_j, $\hat{F}(u) = \{\#(U_j \le u)\}/m'$ where u varies continuously over the interval $(0, 1)$. It is a straightforward, if slightly fiddly, exercise to show that this corresponds precisely to the statistic D defined at eqn. (4.3.4).

4.4 The fast Fourier transform

When the theory of spectral analysis was first established, electronic computers were in a somewhat primitive state of development, and direct evaluation of the periodogram using eqn. (4.2.1) was a formidable task – certainly much more time-consuming than evaluation of the correlogram, for which only the first few autocorrelation coefficients are typically of interest. In a modern computing environment, evaluation of the periodogram is an entirely straightforward exercise for series of the length encountered in most biological investigations – say n of the order of a few hundred at most. Nevertheless, there are dramatic gains in efficiency to be had by using an ingenious 'fast Fourier transform' algorithm (henceforth, FFT), and this becomes important if either very long series are available for analysis or computing is expensive.

The FFT was introduced by Cooley and Tukey (1965), and is most easily explained using the complex number representation of the periodogram,

$$I(\omega) = n^{-1} |d(\omega)|^2, \qquad (4.4.1)$$

where

$$d(\omega) = \sum_{t=0}^{n-1} y_t \exp(it\omega) \qquad (4.4.2)$$

and ω is of the form $\omega_j = 2\pi j/n$ for j an integer between 0 and $n - 1$. Labelling the data as $y_0, \ldots , y_{n-1}$ rather than $y_1, \ldots , y_n$ is no more than a technical device which makes the explanation of the FFT slightly easier without affecting eqn. (4.4.1). The same goes for the apparently redundant specification of frequencies ω in the range 0 to 2π, rather than 0 to π.

Note first that direct evaluation of $d(\omega)$ as defined at eqn. (4.4.2) requires n complex multiplications and additions for each frequency ω, so that the computational cost of evaluating all $m \simeq n/2$ periodogram ordinates is proportional to n^2. However, suppose that we can factorize n as $n = rs$. Then, each t in the range 0 to $n-1$ can be written in the form

$$t = t_0 + rt_1,$$

where t_0 ranges from 0 to $r-1$ and t_1 from 0 to $s-1$. Similarly, each j in $\omega_j = 2\pi j/n$ can be written as

$$j = j_0 + sj_1,$$

where now j_0 ranges from 0 to $s-1$ and j_1 from 0 to $r-1$.

Now, eqn. (4.4.2) can be written as

$$
\begin{aligned}
d(\omega_j) &= \sum_{t_0=0}^{r-1} \sum_{t_1=0}^{s-1} y_{t_0+rt_1} \exp\{2\pi i j(t_0 + rt_1)/(rs)\} \\
&= \sum_{t_0} \exp(2\pi i j t_0/n) \sum_{t_1} y_{t_0+rt_1} \exp(2\pi i j t_1/s) \qquad (4.4.3)
\end{aligned}
$$

Moreover, writing $j = j_0 + sj_1$ gives

$$
\begin{aligned}
\exp(2\pi i j t_1/s) &= \exp\{2\pi i (j_0 + sj_1)t_1/s\} \\
&= \exp(2\pi i j_0 t_1/s) \cdot \exp(2\pi i j_1 t_1) \\
&= \exp(2\pi i j_0 t_1/s),
\end{aligned}
$$

since $j_1 t_1$ is an integer and $\exp(2\pi i) = 1$. It follows that

$$
\begin{aligned}
d(\omega_j) &= \sum_{t_0=0}^{r-1} \exp(2\pi i j t_0/n) \sum_{t_1=0}^{s-1} y_{t_0+rt_1} \exp(2\pi i j_0 t_1/s) \\
&= \sum_{t_0=0}^{r-1} S(t_0, j_0) \exp(2\pi i j t_0/n). \qquad (4.4.4)
\end{aligned}
$$

In eqn. (4.4.4), each $S(t_0, j_0)$ has the same form as eqn. (4.4.2), but for a series of length s rather than n. The computational cost of evaluating each $S(t_0, j_0)$ is therefore proportional to s. Since there are rs distinct combinations of t_0 and j_0, this gives a total cost proportional to $rs^2 = ns$. Finally, eqn. (4.4.4) itself also has the same form as eqn. (4.4.2) but for a series of length r, and must be evaluated for all n values of j, giving an additional cost proportional to nr. Putting these results together, the cost of evaluating all of the $d(\omega)$ indirectly using eqn. (4.4.4) is proportional to $n(r+s)$, which is substantially less than the cost proportional to n^2 entailed by direct evaluation using eqn. (4.4.2). For example, if $n = 100 = 10 \times 10$, $n(r+s) = 2000$ compared with $n^2 = 10\,000$.

Further savings are possible if n can be further factorized, since the same principle can be applied to the valuation of each $S(t_0, j_0)$. Thus, if $n = rs_1s_2 \cdots s_k$, the computational cost can be made proportional to $n(r + s_1 + \cdots + s_k)$. For example, taking $n = 100$ as before, but factorizing 100 as $5 \times 5 \times 2 \times 2$, the cost factor comes down to $100 \times (5 + 5 + 2 + 2) = 1400$.

Early versions of the FFT algorithm required n to be a power of 2, but versions for general n are now widely available, for example in the NAG and IMSL subroutine libraries.

4.5 Periodogram averages

The inconsistency of the periodogram as an estimate of the spectrum stems from its attempting to estimate $m \simeq n/2$ separate quantities $f(\omega_j)$ from n observations y_t. Notice, however, that as n increases, the Fourier frequencies $\omega_j = 2\pi j/n$ constitute an increasingly fine partitioning of the fixed frequency range of interest, namely $0 \leq \omega \leq \pi$. If we are prepared to assume that the underlying spectrum $f(\omega)$ is a smooth function, it follows that $f(\omega)$ is approximately constant within a small interval, $(\omega - \varepsilon, \omega + \varepsilon)$ say, and it is then sensible to estimate $f(\omega)$ by taking the average of all periodogram ordinates $I(\omega_j)$ such that ω_j lies within ε of ω. But the form of ω_j implies that the number of such $I(\omega_j)$ is proportional to n and, because the $I(\omega_j)$ are uncorrelated, the variance of their average is proportional to n^{-1}. In other words, by averaging periodogram ordinates in this way, we gain precision as n increases, at the expense of not gaining resolution of the frequency range.

The above discussion leads us to define an estimator for the spectrum known as a *discrete spectral average of order* $2p + 1$,

$$\hat{f}(\omega_j) = (2p + 1)^{-1} \sum_{l=-p}^{p} I(\omega_{j+l}). \tag{4.5.1}$$

Note that this is precisely a simple moving average, as defined in Section 2.2.1, but applied to the sequence of periodogram ordinates. The sampling distribution of $\hat{f}(\omega_j)$ follows immediately from Theorem 4.1 and the additive property of independent chi-squared variates:

Theorem 4.2. *Let* $\{Y_t\}$ *be a stationary random process with spectrum* $f(\omega)$, $\{y_t : t = 1, \ldots, n\}$ *a partial realization of* $\{Y_t\}$ *and* $I(\omega)$ *the periodogram of* $\{y_t\}$. *Let* $\omega_j = 2\pi j/n$ *for positive integers* $j < n/2$, *and define*

$$\hat{f}(\omega_j) = (2p + 1)^{-1} \sum_{l=-p}^{p} I(\omega_{j+l})$$

for some positive integer p. Then, in the limit n → ∞,

(a) $\hat{f}(\omega_j) \sim f(\omega_j)\chi^2_{2(2p+1)}/\{2(2p+1)\}$
(b) $\hat{f}(\omega_j)$ *and* $\hat{f}(\omega_k)$ *are independent for all j,k such that* $j - k \geqslant 2p + 1$.

Notice that we have stated Theorem 4.2 in terms of a *fixed p*, whereas the motivation for the estimator $\hat{f}(\omega)$ was in terms of p proportional to n. A careful discussion of the optimal choice of p, and its consequences for the asymptotic distribution of the complete set of $\hat{f}(\omega_j)$, involves several delicate issues. From a theoretical point of view, in order to make the sampling variance of $\hat{f}(\omega_j)$ tend to zero as n tends to infinity, it is necessary only that $p \to \infty$, but quite possibly much more slowly than n. The advantage of a slower rate of increase is that it reduces the amount of bias introduced by the averaging of successive periodogram ordinates, which in turn is closely connected with the sharpness of any peaks in $f(\omega)$. From a practical point of view, Theorem 4.2 forms a basis for making inferences about $f(\omega)$ using any given $\hat{f}(\omega)$ of the form (4.5.1), but it says nothing about how to choose p. This choice, as in many other smoothing problems, must involve an element of subjective judgement unless we have some prior knowledge of the general shape of $f(\omega)$, in particular, knowledge of the sharpness and separation in the frequency domain of distinct peaks. For example, if the spectrum contains a very sharp peak, extending over only one or two adjacent Fourier frequencies, then any non-zero p will introduce a substantial negative bias into $\hat{f}(\omega)$ at the frequency ω corresponding to the peak. Conversely, if the spectrum is approximately constant over a large number of consecutive Fourier frequencies, then a large value of p will introduce very little bias whilst giving a substantial reduction in the variance of $\hat{f}(\omega)$.

Before proceeding to some examples, we should perhaps reinforce the dual interpretation of a spectral estimate $\hat{f}(\omega)$. On the one hand, $\hat{f}(\omega)$ is an estimate of the spectrum $f(\omega)$ of an underlying stationary random process which is presumed to have generated the data. At this level, we can think of $\hat{f}(\omega)$ simply as a 'signature' of the data without worrying too much about its physical significance.

At a more basic level, we can interpret the set of numbers $\hat{f}(\omega_j): j = 1, \ldots, m$ as a decomposition of the variation in the data into harmonic components without explicit reference to an underlying $f(\omega)$. At this level, peaks in $\hat{f}(\omega)$ correspond to potentially important cyclic patterns of variation. To interpret these cyclic patterns it is often helpful to translate each Fourier frequency ω_j into its corresponding cycle length, this being a rather more tangible concept. The easiest way to do this is to define the total time-span of the data to be $T = n\delta$, where n is the length of the series and δ the time interval between successive observations, and to note that the jth Fourier frequency $\omega_j = 2\pi j/n$ corresponds to j complete cycles in time T.

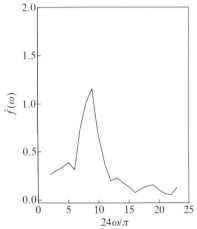

Fig. 4.2. A spectral estimate for the first of the three LH time series, computed as a discrete spectral average of order three.

Example 4.3. Spectral estimates for one LH series
Figure 4.2 shows a spectral estimate for the first of the three LH series in Example 1.2. The estimate is a discrete spectral average of order 3, i.e. $p = 1$ in eqn. (4.5.1). Note that these estimates $\hat{f}(\omega_j)$ are available only at Fourier frequencies $\omega_j = \pi/24$ for $j = 2, \ldots, 23$. It is instructive to compare Fig. 4.2 with the unsmoothed periodogram of these data shown in Fig. 2.15. We see that the main feature of the periodogram, the large peak around the eighth or ninth Fourier frequency, is preserved in the discrete spectral average whilst the erratic fluctuations at other frequencies are substantially reduced.

We can now use Theorem 4.2 to set confidence limits on our estimates $\hat{f}(\omega_j)$. According to the theorem, the sampling distribution of each $\hat{f}(\omega_j)$ is approximately proportional to chi-squared on 6 degrees of freedom,

$$\hat{f}(\omega_j) \sim f(\omega_j)\chi_6^2/6. \tag{4.5.2}$$

To set pointwise confidence limits on the $f(\omega_j)$ we therefore take c_1 and c_2 to be the lower and upper α-critical values of χ_6^2, i.e. $P\{\chi_6^2 \le c_1\} = P\{\chi_6^2 > c_2\} = \alpha$, and evaluate lower and upper confidence limits as

$$l_j = 6\hat{f}(\omega_j)/c_2, \tag{4.5.3}$$

and

$$u_j = 6\hat{f}(\omega_j)/c_1, \tag{4.5.4}$$

respectively. Note that in this example, (4.5.2) holds only for $j = 2, \ldots, 22$. For $j = 23$, the sampling distribution is proportional to χ_5^2

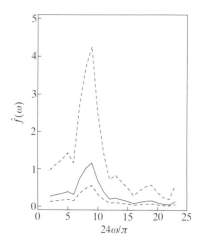 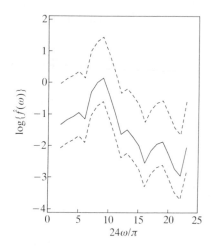

Fig. 4.3. The spectral estimate for the first of the three LH time series and pointwise 90% confidence limits. ———, Spectral estimate; – – –, confidence limits.

Fig. 4.4. The spectral estimate for the first of the three LH time series and pointwise 90% confidence limits, on a logarithmic scale. ———, Spectral estimate; – – –, confidence limits.

rather than χ_6^2, because the sampling distribution of the periodogram ordinate at frequency $\omega_{24} = \pi$ is proportional to χ_1^2 rather than χ_2^2.

Figure 4.3 shows the spectral estimate with pointwise 90% confidence limits. The width of the resulting pointwise confidence intervals serves to warn against too precise an interpretation of the estimated spectrum, although the major peak centred on the ninth Fourier frequency does appear to be a genuine feature. Since the data consist of $n = 48$ values at intervals of $\delta = 10$ minutes, the total time-span of the data is $T = n\delta = 480$ minutes, and the ninth Fourier frequency corresponds to a cycle length of $480/9 \simeq 54$ minutes. Endocrinologists often interpret temporal variation in LH levels in terms of a biological model in which the hormone is released into the bloodstream in a pulsatile fashion (Lincoln *et al.* 1985). Under this model, the characteristic frequency of pulsation is an important parameter to be estimated.

In Fig. 4.3, the confidence intervals for $f(\omega)$ are wider at frequencies for which the estimates $\hat{f}(\omega_j)$ are themselves large. This is a direct consequence of the algebraic form of the limits l_j and u_j given in eqns.

(4.5.3) and (4.5.4). We see that $u_j - l_j$ is proportional to $f(\omega_j)$, whereas

$$\log(u_j) - \log(l_j) = \log(c_2) - \log(c_1) \qquad (4.5.5)$$

does not depend on j, or indeed on any other aspect of the data, but only on the order, p, of the discrete spectral average. In part because of this, spectral estimates are often plotted on a logarithmic scale, interpretation of the statistical fluctuations in the estimates then being much more straightforward. This is particularly the case when we need to compare estimates from several series.

Example 4.3 (continued)
Figure 4.4 reproduces the information in Fig. 4.3 but on a logarithmic scale, confirming that the confidence limits on the logarithmic scale take the form of two parallel lines (except at ω_{23}, for the reason given in the discussion of Example 4.3, although this is not obvious on the diagram).

Example 4.4. Spectral estimates for three LH series
Recall that the three LH series in Example 1.2 consist of one series from the early follicular phase of the subject's menstrual cycle and two series from the late follicular phase of two successive cycles. Figure 4.5 shows log-transformed spectral estimates for the three series, again using a

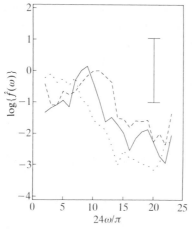

Fig. 4.5. Spectral estimates for the three LH time series, each computed as a discrete spectral average of order three and shown on a logarithmic scale. The vertical bar represents the width of pointwise 90% confidence limits at each frequency. ———, Late follicular phase of first menstrual cycle; - - - -, early follicular phase of second menstrual cycle; – – –, late follicular phase of second menstrual cycle.

discrete spectral average of order 3 in each case. The (constant) width of the 90% pointwise confidence intervals is shown as a vertical bar on the diagram. This suggests that the differences between the spectral estimates for the two late follicular series are certainly within the limits of statistical fluctuation. The estimate for the early follicular series is somewhat different to the other two in that it does not show a clear peak, but decreases steadily as the frequency increases. The width of the pointwise confidence intervals suggests that this apparent difference may again be within the limits of statistical fluctuation, although the evidence is less clear-cut. We shall pursue this point in Section 4.8.

4.6 Other smooth estimates of the spectrum

4.6.1 Weighted moving averages

A natural generalization of the discrete spectral average estimator (4.5.1) is a *weighted* moving average of periodogram ordinates,

$$\hat{f}_w(\omega_j) = \sum_{l=-p}^{p} w_l I(\omega_{j+l}), \qquad (4.6.1)$$

where $w_l = w_{-l}$, $w_0 \geq w_1 \geq \cdots \geq w_p > 0$ and $\sum_{l=-p}^{p} w_l = 1$. The sampling distribution of $\hat{f}_w(\omega)$ is approximately proportional to chi-squared on ν degrees of freedom, where ν, called the 'equivalent degrees of freedom' is chosen so as to match the approximate mean and variance of the estimator $\hat{f}_w(\omega)$ and the approximating chi-squared distribution. Note firstly that

$$E[\hat{f}_w(\omega)] = \sum_{l=-p}^{p} w_l f(\omega_{j+l})$$

$$\simeq f(\omega_j) \sum_{l=-p}^{p} w_{j+l} = f(\omega_j),$$

from which we deduce that the chi-squared approximation must take the form

$$\hat{f}_w(\omega) \sim f(\omega)\chi_\nu^2/\nu. \qquad (4.6.2)$$

Now, to identify ν, note that

$$\text{var}\{\hat{f}_w(\omega_j)\} = \sum_{l=-p}^{p} w_l^2 \{f(\omega_{j+l})\}^2$$

$$\simeq \{f(\omega_j)\}^2 \sum_{l=-p}^{p} w_l^2. \qquad (4.6.3)$$

According to (4.6.2), the variance of $\hat{f}_w(\omega_j)$ is

$$\{f(\omega_j)\}^2(2\nu/\nu^2) = 2\{f(\omega_j)\}^2/\nu. \tag{4.6.4}$$

Finally, equating (4.6.3) and (4.6.4) gives

$$\nu = 2 \Big/ \Big\{ \sum_{l=-p}^{p} w_l^2 \Big\}. \tag{4.6.5}$$

Note in particular that for a simple moving average, with $w_l = (2p+1)^{-1}$ for all l, (4.6.5) reduces to $\nu = 2(2p+1)$ in agreement with Theorem 4.2.

Viewed purely as a smoothing exercise, the use of non-uniform weights is intuitively sensible. However, uniform weighting has the advantage that the estimator is a simple average over a fixed band of frequencies, which makes for a more tangible interpretation. Furthermore, as in most other smoothing problems the precise functional form of the weights is much less important than the overall amount of smoothing as determined by the value of p.

4.6.2 Spectral windows

Before the advent of the FFT, the usual way of computing the periodogram was via eqn. (4.2.3):

$$I(\omega) = g_0 + 2 \sum_{k=1}^{n-1} g_k \cos(k\omega). \tag{4.6.6}$$

Looked at in this light, a plausible explanation of the unsatisfactory statistical properties of $I(\omega)$ is that the high-order sample auto-covariances g_k contribute essentially no more than random noise, since in most practical situations the theoretical autocovariances γ_k decay to zero as k increases. This in turn suggests that an improved estimate of the spectrum might be obtained by discarding the high-order g_k to give

$$I_K(\omega) = g_0 + 2 \sum_{k=1}^{K} g_k \cos(k\omega). \tag{4.6.7}$$

Note that this introduces bias if the corresponding γ_k for $k \geq K$ are non-zero. As usual in smoothing problems, we can improve precision at the expense of introducing bias, and *vice versa*.

More generally, we introduce a non-increasing sequence of weights λ_k and define an estimate

$$\hat{f}_\lambda(\omega) = g_0 + 2 \sum_{k=1}^{n-1} \lambda_k g_k \cos(k\omega). \tag{4.6.8}$$

The sequence $\{\lambda_k\}$ is called a *lag window,* following Blackman and Tukey (1959).

It turns out that these 'windowed' estimators for the spectrum are equivalent to the weighted moving average estimators of Section 4.6.1. To see this, write eqn. (4.6.6) as

$$\hat{f}_w(\omega_j) = \sum_{l=-p}^{p} w_l \left[g_0 + 2 \sum_{k=1}^{n-1} g_k \cos\{k(\omega_j + 2\pi l/n)\} \right]$$

$$= g_0 + 2 \sum_{k=1}^{n-1} g_k \left[w_0 \cos(k\omega_j) \right.$$

$$\left. + \sum_{l=1}^{p} w_l \{\cos(k\omega_j + 2\pi kl/n) + \cos(k\omega_j - 2\pi kl/n)\} \right],$$

Because $\sum w_l = 1$ and $w_l = w_{-l}$. Now, expanding the cosine terms according to the formula

$$\cos(\alpha + \beta) = \cos(\alpha)\cos(\beta) - \sin(\alpha)\sin(\beta)$$

gives

$$\hat{f}_w(\omega_j) = g_0 + 2 \sum_{k=1}^{n-1} \lambda_k g_k \cos(k\omega_j)$$

as required, with

$$\lambda_k = w_0 + 2 \sum_{l=1}^{p} w_l \cos(2\pi kl/n). \qquad (4.6.9)$$

Because of this equivalence, and the existence of the FFT, windowed estimates of $f(\omega)$ have little operational importance. However, apart from their historical interest they are of theoretical importance because a considerable amount of effort has been invested into the investigation of the statistical properties of windowed estimates and the optimal choice of the sequence $\{\lambda_k\}$ in relation to the form of the underlying $f(\omega)$. Accordingly, Table 4.1 gives the form of some $\{\lambda_k\}$ which have been so investigated, and a reference to the origin of each. For an overview, see Priestley (1981, Section 6.2.3).

4.6.3 Spline regression estimates

Theorem 4.1 part (a) asserts that the relationship between the periodogram ordinates $I(\omega_j)$ and the corresponding spectral ordinates $f(\omega_j)$ takes the form of a regression,

$$\log\{I(\omega_j)\} = \log\{f(\omega_j)\} + Z_j \qquad (4.6.10)$$

where the random variable Z_j represents the logarithm of half a

Table 4.1. Some possible specifications of $\{\lambda_k\}$ for windowed spectral estimates. In all cases, K controls the amount of smoothing (larger K corresponding to less smoothing)

Name	λ_k	References
Bartlett	$\begin{cases} 1 - k/K & (k \leq K) \\ 0 & (k > K) \end{cases}$	Bartlett (1950)
Daniell	$\{\sin(\pi k/K)\}/(\pi k/K)$	Daniell (1946)
Parzen	$\begin{cases} 1 - 6(k/K)^2 + 6(k/K)^3 & (k \leq K/2) \\ 2(1 - k/K)^3 & (K/2 < k \leq K) \\ 0 & (k > K) \end{cases}$	Parzen (1961)
Tukey–Hanning	$\begin{cases} \{1 + \cos(\pi k/K)\}/2 & (k \leq K) \\ 0 & (k > K) \end{cases}$	Tukey (1949) Blackman and Tukey (1959)

chi-squared variate on two degrees of freedom. Furthermore, part (b) of the same theorem implies that the Z_j are approximately independent random variables. In other words, the statistical relationship between the log-periodogram ordinates, $L(\omega_j) = \log\{I(\omega_j)\}$, and the log-spectrum, $g(\omega) = \log\{f(\omega)\}$, is that of a standard, albeit non-linear, regression model.

The above discussion suggests estimating $g(\omega)$ by a non-parametric regression of the $L(\omega_j)$ on the corresponding ω_j. A convenient way to do this is by spline regression, as discussed in Section 2.2.3. Note also that in view of the approximate independence of the Z_j in eqn. (4.6.10), generalized cross-validation can be used to choose the amount of smoothing of the $L(\omega_j)$ automatically. A possible objection to this form of automatic smoothing is that the cross-validation prescription effectively puts all the log-transformed periodogram ordinates on the same footing, whereas in many practical situations the 'interesting' part of the spectrum is confined to a sub-interval of $(0, \pi)$. For example, the spectrum may show quite sharp peaks and troughs at low frequencies whilst decaying smoothly at higher frequencies; indeed, when we sample from an underlying process in continuous time, we deliberately try to choose a sampling interval which gives precisely this result, indicating that no important frequency components of the underlying process have been lost by the choice of sampling interval. The practical implication of this is that the generalized cross-validation prescription may lead to a very smooth estimate in circumstances where we would prefer to tolerate rough estimates at high frequencies in order to preserve resolution in the more interesting low-frequency range.

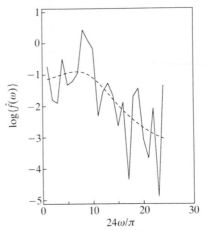

Fig. 4.6. A spline regression estimate of the log spectrum of the first of the three LH time series. ——, Logarithms of periodogram ordinates; – – –, spline regression estimate using generalized cross-validation.

Example 4.5. Spline regression estimate of the spectrum of an LH series
Figure 4.6 compares the log-periodogram for the first of the three LH series in Example 1.2 with the spline regression estimate using generalized cross-validation. Note that this estimate is much smoother than the discrete spectral average of order 3 shown in Fig. 4.4. In particular, the major peak in the discrete spectral average is almost obliterated.

This approach to spectrum estimation has been further developed by Wahba and Wold (1975) and Wahba (1980). One of their refinements to the simple procedure described above is to recognize that the Z_j in eqn. (4.6.10) have a known mean and variance, $E(Z_j) = -0.577$ and $\text{var}(Z_j) = 1.645$.

4.6.4 *Autoregressive spectral estimates*

A final method, which we mention only briefly, is to fit an AR(p) process to the data $\{y_t\}$ and to use the fitted autoregressive spectrum as the estimate of $f(\omega)$. The motivation for this is threefold: fitting an autoregressive process is computationally easy, autoregressive spectra can assume a wide variety of shapes, and automatic criteria are available for choosing the value of p. Nevertheless, the method seems to fit uneasily into a discussion of what is essentially a non-parametric estimation problem. It is analogous to the use of polynomial regression for data smoothing, and is open to the same basic objection, namely that it imposes global assumptions which can lead to artefacts in the estimated spectrum.

4.7 Adjusting spectral estimates for the effects of filtering

In Section 2.6 we pointed out that trend removal from time-series data 'corrupts' the autocorrelation structure of the residual time series, leading to possible misinterpretation. One advantage of spectral analysis over methods based on the correlogram is that, at least when trend-removal is achieved by linear filtering, the theory developed in Section 3.3 gives a simple means of adjusting the estimates to take account of the corruption. There, we showed that if two stationary random sequences $\{R_t\}$ and $\{Y_t\}$ were related according to the equation

$$R_t = \sum_{j=-\infty}^{\infty} a_j Y_{t-j}, \qquad (4.7.1)$$

then the corresponding spectra satisfied the relationship

$$f_r(\omega) = |a(\omega)|^2 f_y(\omega), \qquad (4.7.2)$$

where

$$a(\omega) = \sum_{j=-\infty}^{\infty} a_j e^{-ij\omega} \qquad (4.7.3)$$

is called the *transfer function* of the linear filter (4.7.1).

In the present context, suppose that we smooth a time series $\{y_t\}$ by a linear smoother of the form

$$s_t = \sum_{j=-p}^{p} w_j y_{t-j}. \qquad (4.7.4)$$

Then, except near the ends of the series, the corresponding series of residuals is

$$r_t = y_t - s_t = \sum_{j=-p}^{p} w_j^* y_{t-j}, \qquad (4.7.5)$$

where $w_0^* = 1 - w_0$ and $w_j^* = -w_j$ for all $j \neq 0$. These results express the residual series $\{r_t\}$ and the smooth series $\{s_t\}$ as linear filters of $\{Y_t\}$. If, as is usual, $w_j = w_{-j}$ in eqn. (4.7.4), the corresponding transfer functions to convert $\{Y_t\}$ to $\{r_t\}$ and $\{s_t\}$ respectively are

$$a(\omega) = w_0^* + 2 \sum_{j=1}^{p} w_j^* \cos(j\omega).$$

and

$$b(\omega) = w_0 + 2 \sum_{j=1}^{p} w_j \cos(j\omega).$$

Note that $a(\omega) = 1 - b(\omega)$.

In Example 2.1, we applied two different moving average smoothers to the data of Example 1.5, namely

(a) $p = 1$, $w_j = \frac{1}{3}, \frac{1}{3}, \frac{1}{3}$.
(b) $p = 6$, $w_{-6} = w_6 = \frac{1}{24}$, $w_j = \frac{1}{12}$ otherwise.

There we remarked that using (a) retained a seasonal pattern of variation in the smoothed sequence, whereas using (b) eliminated it. We can now interpret these remarks in terms of the transfer functions $b(\omega)$ associated with (a) and (b).

Example 4.6. Transfer functions for smoothing the bronchitis data
Figure 4.7 shows $\{b(\omega)\}^2$ for each of the two moving averages used on the bronchitis data. Note firstly that both graphs approach zero at high frequencies. From eqn. (4.7.2), we conclude that in each case the effect of the smoothing is to lower the power at high frequencies – in other words, that the moving averages do indeed smooth the data. Secondly, $\{b(\omega)\}^2$ approaches zero much more rapidly for the 13-point moving average (b) than for the 3-point moving average (a) – in other words, the longer moving average is the more drastic smoother. Thirdly, noting that for monthly data seasonal variation corresponds to a frequency $\omega =$

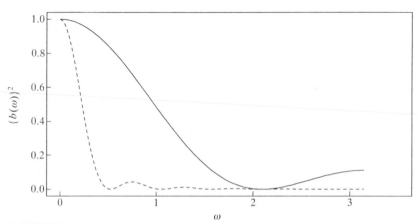

Fig. 4.7. The effect of moving average smoothing on the spectrum of the smoothed series. ——, 3-Point moving average; – – –, 13-point moving average.

$\pi/6 \approx 0.52$, we see that using (b) gives $\{b(\omega)\}^2 \approx 0$ in the vicinity of $\omega = \pi/6$, confirming that (b) eliminates seasonal variation.

We can conduct a similar analysis to elucidate the effect of the moving average smoothing on the spectral properties of the residual series $\{r_t\}$. Firstly, note from eqn. (4.7.2) that if we calculate an estimated spectrum, $\hat{f}_r(\omega)$, say, from the residual series we can convert this to an estimate, $\hat{f}_y(\omega)$, say, of the spectrum of the original process $\{Y_t\}$ by defining

$$\hat{f}_y(\omega) = \hat{f}_r(\omega)/\{a(\omega)\}^2. \tag{4.7.6}$$

Figure 4.8 shows $\{a(\omega)\}^2$ for each of the two moving averages used on the bronchitis data. In both cases, $a(0) = 0$, from which it follows that $\hat{f}_y(0)$ is undefined. Whereas this is not a major problem in itself, it does suggest that any estimate obtained using eqn. (4.7.6) will be unreliable for small ω. Put the other way round, if we are interested in the behaviour of $f_y(\omega)$ at low frequencies we should analyse $\{y_t\}$ itself, rather than $\{r_t\}$.

This last remark brings us to the second, and more important comment on eqn. (4.7.6). In practice, we often use smoothing as a device for converting a non-stationary series $\{y_t\}$ into an approximately stationary series $\{r_t\}$. That is, we assume firstly that the series $\{y_t\}$ is actually generated by a process $Y_t^* = \mu(t) + Y_t$, where $\{Y_t\}$ is stationary, and secondly that subtraction of s_t approximately eliminates $\mu(t)$, leaving r_t as a realization of a linearly filtered version of $\{Y_t\}$ (cf. Section 2.6). However, it remains the case that de-trending by a linear smoother of the form (4.7.4) effectively sacrifices any information about the spectrum of $\{Y_t\}$ at very low frequencies.

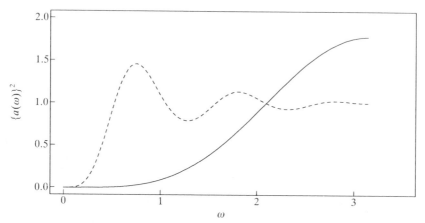

Fig. 4.8. The effect of moving average smoothing on the spectrum of the residual series. ———, 3-Point moving average; – – –, 13-point moving average.

Example 4.7. Estimating the spectrum of bronchitis residuals
In Example 2.8b, we calculated the correlograms of residual series $\{r_t\}$
obtained by subtracting from each of the male and female bronchitis data
an unweighted 3-point moving average. We noted that in each case, the
lag one autocorrelation was negative and speculated that this might be
explainable as by-product of the smoothing. We are now in a better
position to follow up this speculation.

Consider first the male data. In Fig. 4.9a we show a discrete spectral

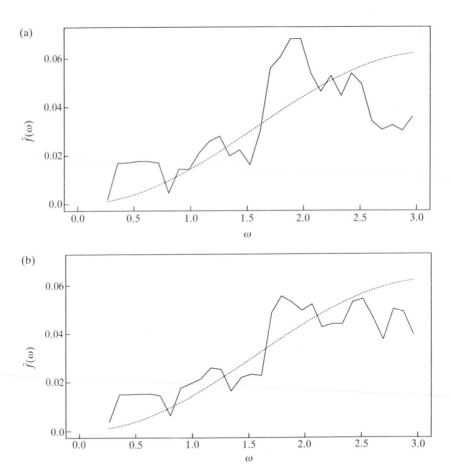

Fig. 4.9. A spectral estimate for the residual series of monthly returns of deaths
in the United Kingdom attributed to bronchitis, emphysema, and asthma. (a)
Males, (b) Females. ——, Spectral estimate, computed as a discrete spectral
average of order five; - - - - -, the function $\{a(\omega)\}^2$.

average of order 5, calculated from the male residuals. The estimate has been scaled so that $\sum \hat{f}(\omega_j) = 1$. Superimposed on $\hat{f}(\omega)$ is the function $\{a(\omega)\}^2$, scaled in the same way. We see a qualitative agreement between the two functions, suggesting that the form of the residual spectrum is largely a by-product of the moving average used to filter the data. This reinforces the view that the original data might consist of a strong seasonal trend plus independent random fluctuations about the trend.

Figure 4.9b tells the corresponding story for the female data. Here, the agreement between the scaled $\hat{f}(\omega)$ and $\{a(\omega)\}^2$ is somewhat better. On the other hand, the erratic fluctuations in $\hat{f}(\omega)$ provide a warning that the data are too sparse to allow a sharp conclusion to be drawn.

4.8 Combining and comparing spectral estimates

Until now, we have assumed that only one time series is available for estimating the spectrum. However, biological investigations usually involve deliberate replication of a basic experiments, as in Examples 1.2, 1.5, 1.6 and 1.7. Suppose, then, that our data consist of a number of time series, one from each of a number of experimental units which may themselves be allocated amongst several different experimental treatments. Two questions immediately arise. How can we estimate a spectrum from a number of replicate series? How can we compare the estimated spectra corresponding to different experimental conditions?

To answer the first question, consider a data set consisting of r replicate series each of length n, and let $I_k(\omega_j)$ denote the periodogram ordinate of the kth series at frequency $\omega_j = 2\pi j/n$. An obvious estimator for the underlying spectrum $f(\omega_j)$ is

$$\bar{I}(\omega_j) = r^{-1} \sum_{k=1}^{r} I_k(\omega_j). \qquad (4.8.1)$$

Two immediate consequences of Theorem 4.1 are that, for large n,

$$\bar{I}(\omega_j) \sim f(\omega_j)\chi_{2r}^2/(2r), \qquad (4.8.2)$$

and that $\bar{I}(\omega_j)$ and $\bar{I}(\omega_l)$ are independent for $l \neq j$. Note that the averaging operation will tend to smooth out the random fluctuations in the individual periodograms. However, if further smoothing is required the techniques described earlier for smoothing the periodogram of a single series can still be used. In particular, simple averaging over adjacent frequencies preserves the chi-squared nature of the sampling distribution but with increased degrees of freedom, exactly as in Section 4.5.

We now move on to the second question, which in its simplest manifestation might involve a comparison between two periodogram averages, $\bar{I}_1(\omega)$ and $\bar{I}_2(\omega)$, say, based on r_1 and r_2 replicate series respectively. The multiplicative form of the sampling distribution in eqn. (4.7.2) suggests examining the set of ratios

$$R(\omega_j) = \bar{I}_1(\omega_j)/\bar{I}_2(\omega_j). \qquad (4.8.3)$$

Then, writing $f_1(\omega)$ and $f_2(\omega)$ for the corresponding two spectra, we deduce that

$$R(\omega_j) \sim \{f_1(\omega_j)/f_2(\omega_j)\}F_{2r_1,2r_2}, \qquad (4.8.4)$$

where $F_{a,b}$ denotes the F-distribution with numerator and denominator degrees of freedom a and b. As in the estimation of a single spectrum, we can apply further smoothing to each of $\bar{I}_1(\omega_j)$ and $\bar{I}_2(\omega_j)$ before computing their ratio $R(\omega_j)$. In particular, if we smooth using the average of $\bar{I}_1(\omega_j)$ over $2p_1 + 1$ adjacent frequencies, and the average of $\bar{I}_2(\omega_j)$ over $2p_2 + 1$ adjacent frequencies, then eqn. (4.8.4) continues to hold, but with numerator and denominator degrees of freedom $2r_1(2p_1 + 1)$ and $2r_2(2p_2 + 1)$, respectively.

The result (4.8.4) can be used to calculate pointwise confidence intervals for the spectral ratios $h(\omega) = f_1(\omega)/f_2(\omega)$, or tolerance intervals for a postulated $h(\omega)$. Note in particular that the hypothesis of a common spectrum in the two treatment groups corresponds to $h(\omega) = 1$ for all ω.

Example 4.8. Comparing spectral estimates for the three LH series
In Example 4.4 Fig. 4.5 we showed spectral estimates for each of three LH series, one from the early follicular phase of the subject's menstrual cycle and two from the late follicular phase of two successive cycles. There, we suggested that the differences between the two late follicular estimates were within the limits of statistical fluctuation. To reinforce this conclusion, Fig. 4.10 shows, on a logarithmic scale, the estimated spectral ratio of the two late follicular series,

$$\hat{R}(\omega) = \hat{f}_1(\omega)/\hat{f}_2(\omega).$$

Because each $\hat{f}_i(\omega)$ is a discrete spectral average of order 3, the sampling distribution of each $\hat{R}(\omega)$ is given by eqn. (4.8.4) with $2r_1 = 2r_2 = 6$,

$$\hat{R}(\omega) \sim h(\omega)F_{6,6},$$

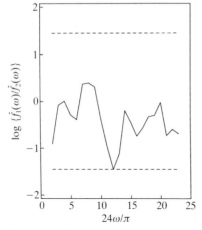 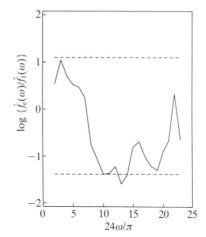

Fig. 4.10. The estimated log spectral ratio for the two late follicular series. ——, Estimate of $\log\{f_1(\omega)/f_2(\omega)\}$; $---$, pointwise 90% tolerance limits assuming $f_1(\omega) = f_2(\omega)$.

Fig. 4.11. The estimated log spectral ratio for the early follicular vs late follicular LH series. ——, Estimate of $\log\{f_e(\omega)/f_l(\omega)\}$; $---$, pointwise 90% tolerance limits assuming $f_e(\omega) = f_l(\omega)$.

where $h(\omega) = f_1(\omega)/f_2(\omega)$. In particular, if c_1 and c_2 denote the lower and upper 5%-critical values of $F_{6,6}$, then the interval (c_1, c_2) is a 90% pointwise tolerance interval for each $\hat{R}(\omega)$ under the assumption that $h(\omega) = 1$, i.e. that the two underlying spectra are the same. These limits are shown as a pair of horizontal dashed lines in Fig. 4.10. All of the estimated ratios $\hat{R}(\omega)$ fall within the tolerance limits, confirming that the differences between the two estimated spectra are interpretable as purely statistical variation in the estimates.

The preceding analysis suggests combining $\hat{f}_1(\omega)$ and $\hat{f}_2(\omega)$ into a single estimate for the late follicular phase,

$$\hat{f}_l(\omega) = \{\hat{f}_1(\omega) + \hat{f}_2(\omega)\}/2.$$

To compare this estimate with the estimate for the early follicular phase, we use the ratios

$$\hat{R}(\omega) = \hat{f}_e(\omega)/\hat{f}_l(\omega),$$

where $\hat{f}_e(\omega)$ is the discrete spectral average of order 3, calculated from the early follicular series. We now repeat the previous analysis using these new ratios, and critical values c_1 and c_2 from $F_{6,12}$. Figure 4.11

shows the result. Now, several of the ratios fall outside the tolerance limits. Moreover, $\log \{R(\omega)\}$ shows a clear trend, suggesting a systematic difference between the underlying spectra in the early and late follicular phases.

Note, incidentally, that in general the *lower* p-critical value of $F_{a,b}$ is the reciprocal of the *upper* p-critical value of $F_{b,a}$.

The results given above do not provide formal tests for the hypothesis that the complete spectrum is the same in two or more treatment groups. For this, we need to combine the information from all the Fourier frequencies ω_j in an appropriate fashion.

In the case of two treatment groups, a procedure analogous to the maximum periodogram ordinate test for white noise is to compute

$$U = \max_{j=1,\ldots,m} \{R(\omega_j)\} \quad \text{and} \quad L = \min_{j=1,\ldots,m} \{R(\omega_j)\}.$$

Under the hypothesis that $h(\omega) = 1$ for all ω,

$$P\{U > u\} = 1 - \{F_0(u)\}^m \tag{4.8.5}$$

and

$$P\{L < l\} = 1 - \{1 - F_0(l)\}^m, \tag{4.8.6}$$

where $F_0(u)$ denotes the probability that an $F_{2r_1, 2r_2}$-distributed random variable assumes a values less than or equal to u. If the attained significance levels (p-values) computed from eqns (4.8.5) and (4.8.6) are p_1 and p_2, then a conservative p-value for a combined test is twice the smaller of p_1 and p_2.

The above test was introduced by Coates and Diggle (1986), who also considered using the ratio U/L to test the hypothesis that $h(\omega)$ is constant, but not necessarily equal to one. This would correspond to equal normalized spectra or, equivalently, a common spectral shape in the two experimental groups.

Tests based on U and L are not particularly powerful, but have the virtue of simplicity in that the values of U and L can be read directly from a graph of the $R(\omega_j)$ against ω_j. A table of selected critical values for the statistic U/L when $r_1 = r_2$ is given in Potscher and Reschenhofer (1988).

Coates and Diggle also considered a semiparametric approach to the problem of comparing estimated spectra, which draws on the ideas of generalized linear modelling (McCullagh and Nelder, 1983) and applies equally to two or more treatment groups.

To avoid unnecessarily cumbersome notation, we shall assume equal replication r in each of t treatment groups, and write I_{kj} for the periodogram average in the kth treatment group at the jth Fourier frequency, $\omega_j = 2\pi j/n$. The extension to unequally replicated treatments requires only adjustment to the degrees of freedom of the chi-squared sampling distributions associated with the I_{kj}.

We first recall from Theorem 4.1 that the I_{kj} are approximately independent random variables with sampling distributions given by

$$I_{kj} \sim f_k(\omega_j) \chi^2_{2r}/(2r),$$

where $f_k(\omega)$ denotes the spectrum in treatment group k. Because the spectrum must be non-negative, we re-write this as

$$I_{kj} = \exp(\theta_{kj}) Z_{kj} \quad : \quad k = 1, \ldots, t; \quad j = 1, \ldots, m, \qquad (4.8.7)$$

where the Z_{kj} are independent χ^2_{2r}-variates and

$$\theta_{kj} = \log\{f_k(\omega_j)/(2r)\}.$$

Equation (4.8.7) defines a *saturated* generalized linear model for the two-way array of periodogram averages, with one parameter θ_{kj} corresponding to each datum I_{kj}. Within this framework, we can obtain testable models by postulating various relationships amongst the θ_{kj}, thereby reducing the number of parameters. For example, the hypothesis of a common spectrum in all t treatment groups is equivalent to

$$\theta_{kj} = \theta_j \quad : \quad j = 1, \ldots, m, \qquad (4.8.8)$$

whereas the hypothesis of a common normalized spectrum is equivalent to

$$\theta_{kj} = \alpha_k + \theta_j. \qquad (4.8.9)$$

Equation (4.8.9) is of particular interest because it identifies the hypothesis of a common spectral shape in all treatment groups as one of no interaction between the row and column factors in the two-way array I_{kj}. In order to test the hypothesis of a common (normalized) spectrum we need to specify a form for this interaction intermediate between equation (4.8.9) and the saturated model (4.8.7). Now, the context in which we are working suggests that the individual spectra, and therefore the ratios between any pair of spectra, are smooth functions of ω. This suggests replacing the constants in eqn. (4.8.9) by smooth functions of ω, and the simplest such functions are low-order polynomials. Arguing along these lines, Coates and Diggle used a quadratic interaction term to give a model of the form

$$\theta_{kj} = (\alpha_k + \beta_k \omega_j + \gamma_k \omega_j^2) + \theta_j. \qquad (4.8.10)$$

Within the ambit of eqn. (4.8.10) we can identify the following three hypotheses:

$$H_0 \quad : \quad \alpha_k = \alpha, \quad \beta_k = \beta, \quad \gamma_k = \gamma,$$
$$H_1 \quad : \quad \beta_k = \beta, \quad \gamma_k = \gamma,$$
$$H_2 \quad : \quad \alpha_k, \beta_k, \gamma_k \text{ arbitrary.}$$

These hypotheses form a nested sequence, and generalized likelihood ratio tests can be used to choose amongst them. To be specific, let L_i denote the maximized log-likelihood for the I_{kj} under H_i. Then,

(a) if H_0 is the natural null hypothesis, test H_0 against H_2 by referring $2(L_2 - L_0)$ to critical values of $\chi^2_{3(t-1)}$.
(b) if H_1 is the natural null hypothesis, test H_1 against H_2 by referring $2(L_2 - L_1)$ to critical values of $\chi^2_{2(t-1)}$.

The log-likelihoods associated with the model defined by eqn. (4.8.7) together with any subsequent prescription for the θ_{kj} can be evaluated and maximized from first principles, but a more convenient method of implementation is through a computer package for generalized linear models, for example GLIM or GENSTAT. In GLIM terminology, the model has gamma errors, scale parameter r^{-1} and a logarithmic link. For detailed discussion of these concepts, see McCullagh and Nelder (1983). Note that that the quadratic formulation (4.8.10) for the interaction is not obligatory, and unless t is large it is both sensible and practicable to plot the logarithms of $\bar{I}_k(\omega)/\bar{I}_l(\omega)$ against ω for all $\frac{1}{2}t(t-1)$ combinations of k and l, to give an assessment of its adequacy. A completely non-parametric, and thereby cautious approach would use the replication within treatment groups to fit the saturated model (4.8.7) as the alternative hypothesis, rather than eqn. (4.8.10) or some other smooth formulation. The price paid for this caution is a loss of sensitivity when something like eqn. (4.8.10) *is* an adequate approximation, because an unnecessarily large number of parameters is then being estimated from relatively sparse data.

Example 4.8 (continued).
Table 4.2a gives the maximized log-likelihoods associated with fitting the model (4.8.10) to the two late follicular series and testing the hypotheses H_0 and H_1. To test H_0 against H_2, we compute $2(L_2 - L_0) = 3.64$ and refer this to critical values of χ^2_3. To test H_1 against H_2 we compute $2(L_2 - L_1) = 2.81$ and refer this to critical values of χ^2_2. Clearly, neither result is significant. This strengthens the justification for combining the two spectral estimates as was done in Fig. 4.11.

The same approach can be used to test whether all three series, i.e. including the series from the early follicular phase, share a common underlying spectrum. The results are given in Table 4.2b. From this table, we compute $2(L_2 - L_0) = 15.78$, refer this to critical values of χ^2_6, and conclude that there are significant differences amongst the three underlying spectra ($p \approx 0.015$). Similarly, we compute $2(L_2 - L_1) = 12.84$, refer this to χ^2_4 and reject the hypothesis of proportional spectra ($p = 0.012$).

Table 4.2. Maximized log-likelihood associated with fitting model (4.8.10) to LH series

Hypothesis	Maximized log-likelihood
(a) Two late follicular series	
H_0	-10.42
H_1	-10.51
H_2	-9.10
(b) All three series	
H_0	23.21
H_1	21.74
H_2	15.32

Our final interpretation of these data is therefore that the underlying spectra in the early and late follicular phases differ in shape. Figure 4.12 shows our estimates of the two spectra together with pointwise 90% confidence limits. For the early follicular phase, the estimate is a discrete spectral average of order 3, and the confidence limits derive from χ_6^2. In the late follicular phase, the estimate is an average of two discrete spectral averages of under 3, and the confidence limits derive from χ_{12}^2.

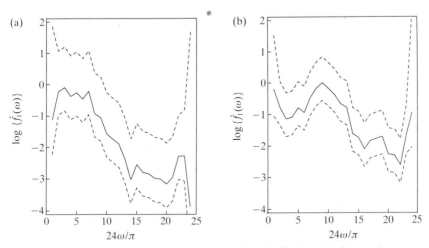

Fig. 4.12. Spectral estimates for the early and late follicular LH time series. ——, Spectral estimate; – – –, pointwise 90% confidence limits. (a) Early follicular phase, spectral estimate a discrete spectral average of order three. (b) Late follicular phase, spectral estimate the average of two discrete spectral averages of order three.

Note once more that by plotting spectral estimates on a logarithmic scale we ensure that the confidence limits appear as parallel lines, except at the two extreme Fourier frequencies where there has necessarily been less smoothing of the periodogram ordinates. The peak in the late follicular spectrum is centred on the ninth Fourier frequency, corresponding to nine complete cycles during the eight-hour span of the data (48 observations at 10 minute intervals), or very roughly one cycle per hour.

The generalized linear modelling approach to the comparison of spectra is strikingly simple to apply because it exploits the link between the apparently esoteric concept of a spectrum and the reassuringly familiar one of multiple regression. One important practical limitation is the assumption that the time series to be analysed are all of the same length. To a limited extent, the methods can cope with unequal length series if the numbers involved factorize neatly. For example, if we wish to compare series A and B of lengths n and kn respectively, we can use as the basis for comparison the complete periodogram of series A and every kth periodogram ordinate of series B. In both cases, the frequencies involved are of the form $\omega_j = 2\pi j/n$. If we are prepared to smooth the periodogram ordinate of series B over adjacent frequencies we can preserve the required correspondence between the two sets of ω_j whilst increasing the precision of the spectral estimate for series B. This discussion presumes that the sampling interval between successive observations remains the same in all series.

4.9 Fitting parametric models

The connection with generalized linear models can also be exploited to fit parametric models to individual periodograms. These ideas have been developed by Cameron and Turner (1987). The starting point is again to express the periodogram ordinates in the form of a multiplicative model,

$$I(\omega_j) = f(\omega_j)Z_j/2 \quad : \quad j = 1, \ldots, m,$$

where the Z_j are mutually independent chi-squared variates on two degrees of freedom, and then to parametrize the spectrum $f(\omega)$.

A particularly simple example of this approach is provided by Bloomfield's (1973) class of models, in which

$$f(\omega) = \exp\left\{2 \sum_{k=0}^{p} \theta_k \cos(k\omega)\right\}, \tag{4.9.1}$$

where p denotes the 'order' of the model. This model has much the same status as the autoregressive process of order p in that it provides a flexible yet tractable class of empirical models for describing stationary time-series data.

In order to express eqn. (4.9.1) in the generalized linear model framework, let $\bar{I}_j$ denote the periodogram average, from r replicates, say, at frequency ω_j, and define an m by $(p + 1)$ matrix X with elements

$$x_{jk} = \cos(k\omega_j)/r$$

Then, if $\boldsymbol{\theta}$ denotes the $(p + 1)$-element vector $(\theta_0, \theta_1, \ldots, \theta_p)'$ and $\boldsymbol{\eta} = X\boldsymbol{\theta}$, it follows that

$$\bar{I}_j = \exp(\boldsymbol{\eta})Z_j,$$

where the Z_j are approximately independent chi-squared variates on $2r$ degrees of freedom. Thus, as in Section 4.8, we have a generalized linear model with gamma errors and logarithmic link.

One methodological advantage of this approach to model-fitting is that attention can be restricted to a subset of the Fourier frequencies if this is sensible in the particular practical context. Cameron and Turner give an example in which the data are an EEG record sampled at a rate of 300 observations per second and the periodogram contains a spike at the Fourier frequency corresponding to the 50 Hz oscillation in the electricity supply. The fit of an autoregressive model for the data is substantially improved by treating the corresponding periodogram ordinate as a missing value.

Example 4.9. Fitting a parametric model to the late follicular LH series
If we look back at Example 4.8, and in particular at Figs 4.11 and 4.12, we can summarize our conclusions as follows:

(a) the underlying spectrum for the early follicular phase shows a steady reduction in power with increasing frequency,
(b) the underlying spectrum for the late follicular phase shows a peak around an intermediate frequency corresponding to an approximate hourly cycle,
(c) the logarithm of the ratio of the early and late follicular spectra is, approximately, a quadratic function of frequency.

The simplest parametric model consistent with (a) is an AR(1) process with positive autoregressive parameter α, i.e. a spectrum of the form

$$f_e(\omega) = \sigma^2/\{1 - 2\alpha \cos(\omega) + \alpha^2\} \qquad (4.9.2)$$

(cf. Example 3.1). In conjunction with (c), this suggests the following parametric model for the late follicular spectrum:

$$f_l(\omega) = f_e(\omega) \exp(a + b\omega + c\omega^2). \qquad (4.9.3)$$

As described by Cameron and Turner (1978), this parametric model can be fitted using the standard machinery of a regression package such as GLIM.

Write $\hat{f}_l(\omega_j)$ for the non-parametric estimate of $f_l(\omega_j)$ obtained by averaging the periodogram ordinates at frequency ω_j from the two late follicular series. The approximate sampling distribution of these quantities follows the now familiar form: the $\hat{f}_l(\omega_j)$ are independently distributed as

$$\hat{f}_l(\omega_j) \sim f_l(\omega_j) \cdot \chi_4^2/4. \tag{4.9.4}$$

For any given value of α, the combination of eqns (4.9.2), (4.9.3), and (4.9.4) defines a generalized linear model with (in GLIM terminology) gamma errors, scale parameter $\frac{1}{2}$ and a logarithmic link. Using the package, it is a straightforward exercise to fit the model for a range of trial values and in this way find the value $\hat{\alpha}$ which maximizes the resulting log-likelihood.

Using this approach leads to estimates $\hat{\alpha} = 0.76$, $\hat{a} = 0.027$, $\hat{b} = -1.078$, and $\hat{c} = 2.494$. Note that the value of $\hat{a}$ is obtained by arbitrarily setting $\sigma^2 = 1$ in eqn. (4.8.1), since a and σ^2 cannot be estimated separately. Figure 4.13 compares the fitted spectrum with the non-parametric estimates $\hat{f}_l(\omega_j)$ and pointwise 90% tolerance limits under the fitted model. All but one of the $\hat{f}_l(\omega_j)$ fall within the tolerance limits, indicating a satisfactory fit.

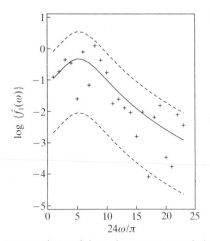

Fig. 4.13. Fitting a parametric model to the spectrum of the late follicular LH time series. +, Non-parametric spectral estimate computed as the average of the periodogram ordinates at each frequency. ——, Fitted spectrum; – – –, pointwise 90% tolerance limits for the non-parametric estimate, assuming the fitted spectrum.

4.10   Discussion: strengths and weaknesses of spectral analysis

In this chapter we have developed a range of methods for the analysis of time-series data. The common basis for these methods is a decomposition of the variation in a time series of length n into harmonic components at the Fourier frequencies $\omega_j = 2\pi j/n : j = 1, \ldots, n/2$. An important consideration for the potential user of spectral methods is whether such a decomposition is readily interpretable. It has to be acknowledged that many classically trained statisticians do not find it so, and prefer the (statistically) more familiar concept of correlation as a descriptor of serial dependence in a time series. In contrast, electronic engineers naturally think of a 'signal' as consisting of a superposition of components at various electrical frequencies. In short, the question of interpretability is to some extent a matter of personal taste. However, it is at least arguable that in biology, many phenomena exhibit patterns of more or less cyclic variation, and give rise to precisely the type of data for which spectral methods are well suited.

On a methodological point, one great strength of spectral methods is the approximate independence of periodogram ordinates at the Fourier frequencies. This leads to technically very straightforward methods of statistical analysis to answer a wide range of questions, as in Sections 4.8 and 4.9. Furthermore, it can be shown that the resulting methods are not critically dependent on the underlying random sequences being Normally distributed (Hannan 1973).

The restriction of spectral methods to stationary phenomena is of central importance. The presumption is that any non-zero mean value, whether constant or time-varying, is either absent or can be removed by de-trending prior to analysis. If a description of trend, whether in isolation or in relation to different experimental treatments, is of direct interest, spectral methods are inappropriate.

Two practical limitations on the interpretation of a periodogram are those which arise from 'harmonics' and from 'leakage'. These are best explained by way of examples.

Example 4.10. Harmonics
Figure 4.14a shows a time series of 100 observations, consisting of the sequence $1, 2, \ldots, 10$ repeated 10 times. Clearly, this is a cylically repeating pattern with period 10, or frequency $\omega_0 = \pi/5$. Figure 4.15 shows the periodogram of the series. As would be expected, the periodogram has a large spike at frequency ω_0, but also smaller ones at $2\omega_0$, $3\omega_0$, $4\omega_0$ and $5\omega_0$. The integer multiples of ω_0 are called its harmonics, and the secondary spikes at these higher frequencies arise because the cyclic variation in the original series is non-sinusoidal. In

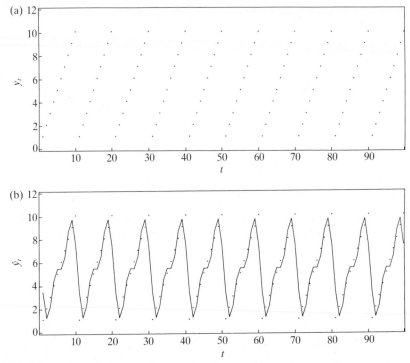

Fig. 4.14. A synthetic time series of 100 observations. (a) The series of values y_t. (b) The series y_t together with fitted values $\hat{y}_t$ computed as the superposition of two sinusoids.

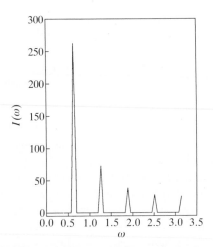

Fig. 4.15. The periodogram of the synthetic time series.

fact, the periodogram ordinates at ω_0 and $2\omega_0$ together account for about 79% of the variation in the original series. Figure 4.14b compares the data y_t with a superposition of sine waves at these two frequencies,

$$\hat{y}_t = \alpha_1 \cos(\omega_0 t) + \beta_1 \sin(\omega_0 t) + \alpha_2 \cos(2\omega_0 t) + \beta_2 \sin(2\omega_0 t),$$

with the α_i and β_i estimated by least squares. It is quite striking that such an obviously non-sinusoidal wave form can be well approximated by the sum of two sine waves. However, the main point of the example is to warn against a naïve interpretation of multiple peaks in the periodogram of a time series as indicating the presence of several distinct cyclic mechanisms in the underlying process.

Example 4.11. Leakage
Leakage refers to the fact that the periodogram divides all the variation in an observed time series amongst a discrete set of Fourier frequencies $\omega_j = 2\pi j/n$, whereas the process generating the data may incorporate a cyclic mechanism at a non-Fourier frequency ω. Figure 4.16a shows a simulated series $\{y_t\}$ of length $n = 100$ generated by a second-order autoregressive process,

$$Y_t = Y_{t-1} - 0.5Y_{t-2} + Z_t,$$

where $\{Z_t\}$ is Normally distributed white noise with variance $\sigma^2 = 1.0$. Figure 4.16 shows the normalized periodogram of this series, together with the theoretical spectrum of the process $\{Y_t\}$, which has a maximum at frequency $\omega_0 = 0.723$. For comparison, Fig. 4.17a shows a simulated realization $\{w_t\}$ generated by a harmonic regression model

$$X_t = 0.8 \cos(\omega_0 t) + Z_t, \qquad (4.10.1)$$

and Fig. 4.17b the normalized periodogram of $\{x_t\}$. In this case, there is no theoretical spectrum because the process $\{x_t\}$ is non-stationary. Nevertheless, the normalized periodograms of $\{y_t\}$ and $\{x_t\}$ look rather similar; in both cases, there is a dominant peak in the periodogram, with moderately large values at nearby frequencies. This is because the variation in the second series $\{x_t\}$, which properly should be ascribed to the frequency ω_0 has been divided, or 'leaked' amongst the Fourier frequencies close to ω_0. Contrast this with Fig. 4.18, which relates to a third series, also generated by the harmonic regression model (4.10.1) except that now $\omega_0 = 0.691$, which is one of the Fourier frequencies for a series of length 100. The dominant peak in the periodogram, which is at frequency ω_0 as expected, is now very sharp. In particular, there is no appreciable leakage into the nearby frequencies; rather, the variation in this third series not ascribed to the frequency ω_0 is distributed more or less evenly throughout the remaining 49 Fourier frequencies.

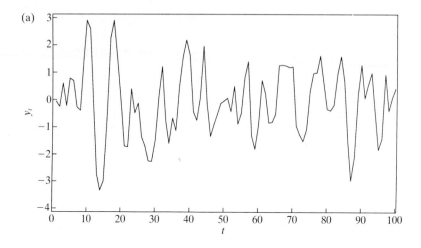

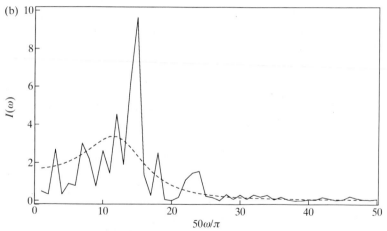

Fig. 4.16. A simulated realization of an AR(2) process, $y_t = y_{t-1} - 0.5Y_{t-2} + Z_t$, its periodogram and spectrum. (a) Simulated realization. (b) ——, Periodogram; $---$, AR(2) spectrum.

Problems of leakage are particularly acute with short series, for which the Fourier frequencies are relatively coarsely spaced. As the length of the series increases, the Fourier frequencies constitute a progressively finer partition of the fixed interval 0 to π, and there is less opportunity for important cyclic components to be sufficiently far from the closest Fourier frequency to cause trouble. Note that leakage is quite distinct from the problem of aliasing which arises because we cannot identify very

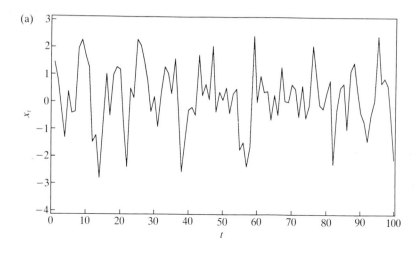

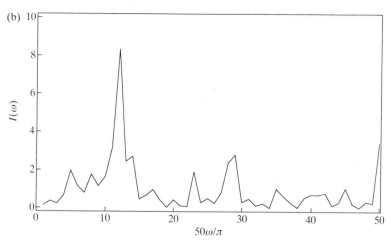

Fig. 4.17. A simulated realization of a harmonic regression model, $W_t = 0.8\cos(0.723t) + Z_t$, and its periodogram. (a) Simulated realization. (b) Periodogram.

high frequency variation from observations at discrete time points (cf. Section 3.2).

As a final comment on spectral analysis, we remark that it allows substantial progress to be made without specifying a parametric model for the data: cases in point are the setting of confidence limits on an estimated spectrum and the contruction of significance tests to compare

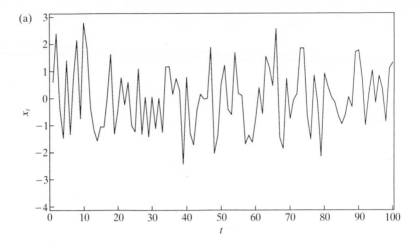

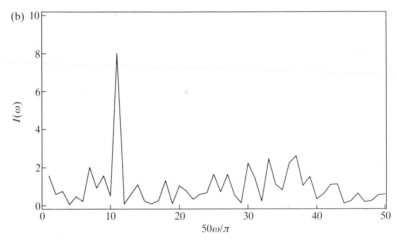

Fig. 4.18. A simulated realization of a harmonic regression model, $W_t = 0.8\cos(0.691t) + Z_t$, and its periodogram. (a) simulated realization. (b) Periodogram.

two or more estimated spectra. In this sense, spectral analysis can be thought of as an essentially non-parametric approach to time-series analysis. As such, we would expect it to perform well when the data are relatively abundant, but to suffer from a lack of sensitivity when the data are sparse. Put more plainly, long series are more suited to spectral analysis than short ones.

To summarize, the following features of a set of time-series data are all positive indicators (and their absence negative indicators) for the successful use of spectral analysis.

- Series are either stationary, or the trend $\mu(t)$ is not of direct interest.
- Series are long.
- Interpretation of underlying processes in terms of cyclic patterns of variation makes sense.
- Parametric models are difficult to justify.

4.11 Further reading

Bloomfield (1976) is a carefully graded introduction to spectral analysis, beginning with simple harmonic regression and emphasising data-analytic methods throughout. Koopmans (1974) gives a more theoretically oriented treatment. Kleiner et al. (1979) demonstrate the sensitivity of periodogram-based methods to outliers in the data, and develop outlier-resistant estimators for the spectrum. Brillinger and Krishnaiah (1983) is a wide-ranging compilation of papers by many different authors.

5
Repeated measurements

5.1 Introduction

In this chapter, we discuss methods for analysing data which consist of a number of relatively short non-stationary time series in which the trends, $\mu_i(t)$ say, are of direct interest. This situation arises in experiments which involve comparisons amongst trends associated with different treatments (cf. Examples 1.6, 1.7). It also applies to growth studies (cf. Example 1.5). We refer to data of this kind as *repeated measurement* data. The distinction between this and the type of replicated time-series data considered in Section 4.8 is not precise in any mathematical sense. Rather, the change of name reflects a change in the practical emphasis of the investigation. In the terminology of Section 1.2, we now seek only to *accommodate* serial dependence within the individual time series in order to make valid inferences about the corresponding $\mu_i(t)$.

In part because of this change in practical emphasis, we shall make the pragmatic assumption that the nature of the random variation is the same in all the individual series. Formally, we assume that the ith series is generated by a random process $\{Y_i(t)\}$ such that

$$Y_i(t) = \mu_i(t) + Z_i(t), \tag{5.1.1}$$

where $\{Z_i(t)\}$ is a *stationary* random process, with the same structure for all series but realized independently for each. In growth studies, where the individual experimental units are a random sample from a single underlying population, we would assume additionally a common trend, $\mu_i(t) = \mu(t)$ for all i. More generally, where the experimental units are allocated amongst several experimental treatments, we assume that treatments affect only the mean response at each time-point, and do not affect the nature of the random variation about the mean response. This is a strong assumption, but analogous to the assumption of constant error variance which underlies classical regression and analysis of variance methodology.

As indicated by the form of eqn. (5.1.1), we shall develop methods for the analysis of repeated measurements in a continuous-time setting, thus automatically embracing observations taken at unequally spaced time points. This again reflects the change in practical emphasis; for example, in growth studies on animals, the experimenter may well take observations relatively often during early, rapid growth, and less frequently as the animals approach maturity. Similarly, we shall not insist that observations be taken at the same set of time points on every series.

In the next section we use the multivariate Normal distribution as a formal framework for repeated measurements. In Section 5.3 we show how time-series structure can be incorporated into this framework, thereby reducing the number of parameters to be estimated. In Section 5.4 we show how the sample variogram can be used to formulate a specific model and provide initial estimates of its parameters. Section 5.5 then develops a procedure for finding maximum likelihood estimates using numerical methods to maximize the likelihood function. Sections 5.2, 5.3, and 5.5 draw heavily on the material in Appendix A. Section 5.6 is a case study based on the data of Example 1.7.

5.2 Repeated measurements as multivariate data

In the model represented by eqn. (5.1.1), suppose that the data consist of sets of observations on different experimental units, each made at the same set of time points $t_j : j = 1, \ldots, l$. Let y_{ij} denote the jth observation on the ith unit, and Y_{ij} the corresponding random variable, i.e. $Y_{ij} = Y_i(t_j)$. Similarly, write $\mu_{ij} = \mu_i(t_j)$ for the mean of Y_{ij}. Now, define l-element vectors $\mathbf{y}_i = (y_{i1}, \ldots, y_{il})'$, $\mathbf{Y}_i = (Y_{i1}, \ldots, Y_{il})'$ and $\boldsymbol{\mu}_i = (\mu_{i1}, \ldots, \mu_{il})'$. Write Σ for the *variance matrix* of $\mathbf{Y}_i$, i.e. Σ is an l by l matrix with (j, k)th element $\sigma_{jk} = \mathrm{cov}(Y_{ij}, Y_{ik})$. Note that we assume a common variance matrix for all $\mathbf{Y}_i$. Finally, assume that each $\mathbf{Y}_i$ has a multivariate Normal distribution. Then, we can express our model for the complete set of data as

$$\mathbf{Y}_i \sim \mathrm{MVN}(\boldsymbol{\mu}_i, \Sigma) \quad : \quad i = 1, \ldots, m. \tag{5.2.1}$$

In order to make (5.2.1) a useful model, we need to impose some structure on either or both of $\boldsymbol{\mu}_i$ and Σ. In the *unstructured multivariate* approach to repeated measurements, we leave Σ entirely unspecified, but use the particular context to suggest an appropriate specification for the $\boldsymbol{\mu}_i$. For example, if the data consist of replicated series on one or more experimental treatments, we would assume that $\boldsymbol{\mu}_i$ depends only on the particular treatment assigned to the ith unit. If, additionally, the data exhibit smooth time trends within each treatment group, we might reflect this in a parametric specification of each $\boldsymbol{\mu}_i$, for example a linear growth model within each treatment group would amount to a specification of the form $\mu_{ij} = \alpha_i + \beta_i t_j$.

Whatever specification we adopt for the $\boldsymbol{\mu}_i$, the multivariate Normal model (5.2.1) involves $\frac{1}{2}l(l+1)$ parameters to define the variance matrix Σ. If l is very much smaller than m, i.e. the data consist of a large number of short series, this does not present any difficulties. We simply estimate the elements of Σ by the sample covariances,

$$\hat{\sigma}_{jk} = m^{-1} \sum_{i=1}^{m} (y_{ij} - \hat{\mu}_{ij})(y_{ik} - \hat{\mu}_{ik}), \qquad (5.2.2)$$

where $\hat{\mu}_{ij}$ denotes an appropriate estimate of μ_{ij}. In particular, if the data consist of m replicate series, then $\mu_{ij} = \mu_j$ does not depend on i and can be estimated by

$$\hat{\mu}_j = m^{-1} \sum_{i=1}^{m} y_{ij}. \qquad (5.2.3)$$

Formal inferential procedures for the multivariate Normal model (5.2.1) are described in most textbooks on multivariate analysis. See, for example, Chatfield and Collins (1980) or, at a more advanced mathematical level, Mardia *et al.* (1979). The practical limitations of this approach are twofold: it is restricted to situations in which observations are made on all experimental units at the same times, and unless l is very much smaller than m it involves estimating an unacceptably large number of parameters. To some extent, the first limitation has been removed by the development of methods for handling missing values in multivariate data (see, for example, Beale and Little 1975 or Crépeau *et al.* 1985). The second limitation is the more fundamental. Indeed, in the context of repeated measurements, data sets with $l > m$ are not uncommon, and in such cases the unstructured multivariate approach fails completely.

Example 5.1. Analysis of data on the body weights of rats
The unstructured multivariate approach can be applied to the data from Example 1.6, in which all 27 rats are observed at the same five time points: $t = 0$, 1, 2, 3 and 4 weeks. In order to estimate the elements of the variance matrix Σ in (5.2.1), we need estimates of the mean value parameters to incorporate into eqn. (5.2.2). The simplest estimates are the fifteen sample means at each of the five time points in each of the three treatment groups. The resulting estimate $\hat{\Sigma}$ is shown below (the elements in the lower triangle are omitted because $\hat{\Sigma}$ is symmetric).

19.92	30.48	29.16	26.05	21.70
	63.44	63.75	55.50	48.28
		87.48	100.98	107.01
			155.10	177.55
				232.80

The first thing to notice about $\hat{\Sigma}$ is that the diagonal elements, representing the variances of observations at the five time points, increase with t. We can also compute estimates of the corresponding correlations, $\hat{\rho}_{jk} = \hat{\sigma}_{jk}/(\hat{\sigma}_{jj}\hat{\sigma}_{kk})$, which are as follows.

1.00	0.86	0.70	0.47	0.32
	1.00	0.86	0.56	0.40
		1.00	0.87	0.75
			1.00	0.93
				1.00

Notice how the correlations decrease with increasing time separation $|j - k|$ between the two measurements concerned, but remain substantially positive throughout. Later, we shall exploit this structure. For the time being, we continue to treat each time sequence of measurements as a multivariate response of dimension five, and make no assumptions about the form of the variance matrix Σ.

The observed mean response profiles in the three treatment groups are as follows:

$$\hat{\mu}_1 = (54.0, 78.5, 106.0, 130.1, 160.6) \quad \text{(control)}$$
$$\hat{\mu}_2 = (55.6, 75.9, 104.9, 132.7, 162.9) \quad \text{(thyroxin)}$$
$$\hat{\mu}_3 = (54.7, 76.3, 95.8, 108.4, 124.2) \quad \text{(thiouracil)}.$$

Multivariate analysis of variance, or MANOVA (Chatfield and Collins 1980, Chapter 8) provides a formal test of the hypothesis that the underlying population mean response profiles are equal, i.e. $\mu_1 = \mu_2 = \mu_3$. Briefly, if $\hat{\Sigma}$ denotes the estimate of the variance matrix given above, and $\hat{\Sigma}_0$ the corresponding estimate using the overall sample means at each of the five time points in eqn. (5.2.2), i.e. assuming $\mu_1 = \mu_2 = \mu_3$, then the likelihood ratio statistic to test the null hypothesis of equal response profiles is

$$D = np\{\log(|\hat{\Sigma}_0|) - \log(|\hat{\Sigma}|)\},$$

where $n = 27$ denotes the number of animals and $p = 5$ the number of observations per animal. Under the null hypothesis, the sampling distribution of D is approximately chi-squared with degrees of freedom $p(k - 1)$, where $k = 3$ denotes the number of different treatments (cf. Section 4.8). For these data, $D = 182.5$, $p(k - 1) = 10$ and we reject the null hypothesis emphatically. We therefore conclude that the mean response profiles differ.

5.3 Incorporating time-series structure

In any statistical modelling exercise, we try to provide an adequate
description of the data using as few parameters as possible. With
repeated measurement data, we can often achieve this by incorporating
time-series structure into the variance matrix of the multivariate Normal
distribution (5.2.1). In so doing we incidentally free ourselves from the
requirement of a single set of observation times common to all series, but
for convenience of exposition we shall not exploit this freedom just yet.

We continue to work within the framework of (5.2.1), whereby the
random vector Y_i of observations on the ith experimental unit has a
multivariate Normal distribution with variance matrix Σ. However, if we
make use of the stationarity assumption in our original model (5.1.1),
then the elements of Σ are constrained to satisfy

$$\sigma_{jk} = \gamma(|t_j - t_k|),$$

where $\gamma(u)$ is the autocovariance function of the zero-mean, stationary
process $Z_i(t) = Y_i(t) - \mu_i(t)$. This immediately reduces the number of
parameters needed to define Σ. For example, if the t_j are equally spaced,
the number of parameters is reduced from $\frac{1}{2}l(l+1)$ to l, because σ_{jk} then
depends only on $|j - k|$.

Example 5.2. Transformation of the data on body weights of rats to
achieve a stationary correlation structure
In Example 5.1, we saw that for the data on body weights of rats, both
the mean response and the variation about the mean response increased
with time. Sometimes, this interrelationship between mean and variance
can be eliminated by a simple transformation of the data (Box and Cox
1964). Below, we give the estimated variance matrix of the *logarithms* of
the body weights, again using the fifteen sample means to estimate the
underlying population mean responses:

0.0068	0.0073	0.0054	0.0041	0.0028
	0.0107	0.0082	0.0065	0.0042
		0.0086	0.0082	0.0070
			0.0103	0.0095
				0.0103

In contrast to the situation for the untransformed data, the variances no
longer increase systematically with time. The corresponding correlation

matrix is

1.00	0.85	0.70	0.49	0.34
	1.00	0.86	0.58	0.41
		1.00	0.88	0.75
			1.00	0.92
				1.00

The correlations again decrease as the corresponding time separation increases; indeed, they agree closely with the correlations calculated from the untransformed data. Furthermore, the correlations at a given time separation show no systematic increase or decrease over time. This suggests that the random variation in the log-transformed data is stationary. Sensible estimates of the stationary autocovariances $\gamma(u)$ are obtained by averaging all of the estimates $\hat{\sigma}_{jk}$ at each value of $u = |j - k|$. Finally, the estimated autocorrelations are obtained as $\hat{\rho}(u) = \hat{\gamma}(u)/\hat{\gamma}(0)$. These are shown graphically in Fig. 5.1, confirming that $\hat{\rho}(u)$ decreases smoothly as the time separation u increases.

Further economies in the number of parameters used to define Σ can be achieved by adopting a parametric model for $\gamma(u)$ itself. What form might such a model take? To be generally useful, a parametric model for repeated measurements should incorporate the following three features. Firstly, the correlation between a pair of observations on the same experimental unit typically decreases as their separation in time increases. This behaviour is evident, for example, in Fig. 5.1. Secondly,

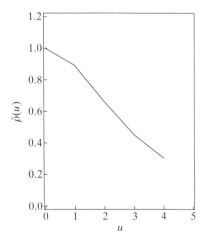

Fig. 5.1. Estimated autocorrelations for the data on log-transformed body weights of rats.

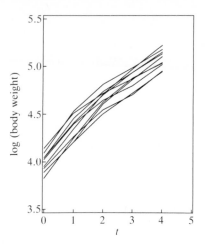

Fig. 5.2. Superimposed time-plots of log-transformed body weights of 10 rats in the control group.

some experimental units may give consistently high responses and others consistently low responses, relative to the overall mean. The same data again illustrate the point. Figure 5.2 shows superimposed time plots for the log-transformed body weights of the ten rats in the control group; note in particular that several of the profiles retain their rank order over the five time points and that the overall spread amongst the ten responses is approximately constant over time. Finally, observations may be subject to random measurement error. This effect should be negligible for the body-weight data, but can be substantial when the measurements in question involve subsampling within each experimental unit as in Example 1.6 on the growth of colonies of *paramecium aurelium*. There, each recorded cell count represents, on average, one tenth of the underlying colony size, but will be subject to random variation about this value.

To express these features in algebraic terms, we decompose the stationary process $\{Z_i(t)\}$ in eqn. (5.1.1) into three components,

$$Z_i(t_j) = \{U_i + V_{ij} + W_i(t_j)\}. \tag{5.3.1}$$

In eqn. (5.3.1), the U_i are mutually independent $N(0, v^2)$ random variables, which represent the variation between experimental units. The V_{ij} are mutually independent $N(0, \tau^2)$ random variables representing measurement error. Finally, the $\{W_i(t)\}$ are independent stationary random processes with common autocovariance function $\sigma^2\rho(u)$ such that $\rho(0) = 1$ and $\rho(u) \to 0$ as $u \to \infty$. For example, we might assume an exponential correlation function,

$$\rho(u) = \exp(-\alpha u),$$

a 'Gaussian' correlation function,

$$\rho(u) = \exp(-\alpha u^2) \tag{5.3.2}$$

or any other appropriate parametric family; we shall discuss the choice of family in Section 5.4.

The autocovariance function $\gamma(u)$ of the process $\{Z_i(t)\}$ defined by eqn. (5.3.1) can readily be derived. Note firstly that

$$\gamma(0) = \text{var}\{Z_i(t)\} = v^2 + \tau^2 + \sigma^2.$$

Now, for any $j \neq k$,

$$
\begin{aligned}
\text{cov}\{Z_i(t_{ij}), Z_i(t_{ik})\} &= E[Z_i(t_{ij})Z_i(t_{ik})] \\
&= E[\{U_i + V_{ij} + W_i(t_{ij})\}\{U_i + V_{ik} + W_i(t_{ik})\}] \\
&= E[U_i^2] + E[W_i(t_{ij})W_i(t_{ik})] \\
&= v^2 + \sigma^2 \rho(|t_{ij} - t_{ik}|).
\end{aligned}
$$

Notice that $\gamma(u) = v^2 + \sigma^2 \rho(u)$ does not approach $\gamma(0)$ as $u \to 0$ because of the presence in the model (5.3.1) of the random variables V_{ij} which represent measurement error and are therefore assumed to be independently distributed even for a pair of coincident t_j. Also, $\gamma(u)$ does not approach zero as $u \to \infty$ because of the presence of the unit-specific random variables U_i. To summarize,

$$
\gamma(u) = \begin{cases} v^2 + \tau^2 + \sigma^2 & : \quad u = 0 \\ v^2 + \sigma^2 \rho(u) & : \quad u > 0. \end{cases} \tag{5.3.3}
$$

Figure 5.3a shows an example of the autocovariance function (5.3.3), assuming the form (5.3.2) for $\rho(u)$ and indicating the contributions from each of the three components of the total variation. Also shown for future reference in Fig. 5.3b is the corresponding variogram, $V(u) = \gamma(0) - \gamma(u)$. We shall use $V(u)$ as a diagnostic tool in preference to $\gamma(u)$ itself because $V(u)$ is more easily estimated from data unequally spaced in time, as discussed in Section 2.5.2.

Until now we have left unspecified the structure of the mean values μ_{ij}. In principle this can be extremely flexible, but in practice analysis is greatly simplified if we confine our attention to *linear* models. These are models in which the vector of means, $\mu_i = (\mu_{i1}, \ldots, \mu_{il})'$, can be expressed in the form

$$\mu_i = X_i \theta, \tag{5.3.4}$$

where θ is a parameter vector of dimension $p \leq l$ and X_i an $l \times p$ matrix

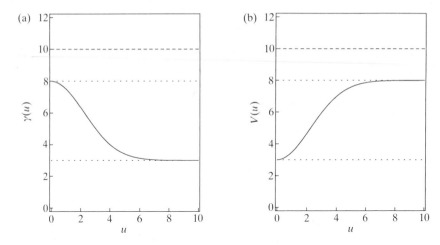

Fig. 5.3. The autocovariance function (5.3.3), with $v^2 = 3$, $\tau^2 = 2$, $\sigma^2 = 5$ and $\rho(u) = \exp(-0.1u^2)$, and the corresponding variogram. (a) Covariance function. (b) Variogram.

with known elements. Note that models of this kind can incorporate treatment effects, covariates and polynomial time trends.

The practical value of this approach to repeated measurements rests on the empirical fact that parametric models of the kind proposed do often provide good approximations to the mean and covariance structure of repeated measurement data. However, this happy state of affairs is by no means universal, and it is therefore important that we develop appropriate diagnostic procedures for formulating a particular model and subsequently checking its adequacy.

5.4 Formulating the model: time plots and the sample variogram

The basic data-analytic tools have already been described in Chapter 2: time plots, smoothing and the variogram. We now consider how these tools can be adapted to the present context, always remembering that we have in mind the analysis of a relatively large number of relatively short series.

Time plots provide an obvious means of elucidating the form of the underlying time trends $\mu_i(t)$. If, as is usual, there is at least some degree of commonality between the times of observation for the different series, it is sensible to plot the average responses from all series within each treatment group against the corresponding observation times. Further smoothing of these averaged time plots may be unnecessary, and in any case is of limited value for very short series.

Time plots can also provide evidence for or against the assumption that the series of deviations about the trends are stationary. For this purpose, it is preferable to construct superimposed time plots of all series *within* each treatment group. If these tend to 'fan out' as time increases, then the variance may be increasing with time. If also the means $\mu_i(t)$ appear to be increasing with time, a variance-stabilizing transformation is often an effective remedy, as was the case for the data on body weights of rats (cf. Example 5.2). When this fails, it may be necessary to abandon the stationarity assumption. Possibilities for further progress then include the unstructured multivariate approach of Section 5.2, or an intermediate stance between this and stationarity.

One example of an intermediate stance in this sense is the approach to growth curve modelling adopted by Sandland and McGilchrist (1979). They argue that the stationarity assumption embodied in eqn. (5.1.1) is more appropriate for the *relative growth rate* (RGR) of an individual. Formally, this leads to a model in which the logarithm of size, $Y(t)$ say, is represented as

$$Y(t) = \mu(t) + \int_0^t W(s)\mathrm{d}s, \tag{5.4.1}$$

where $\{W(t)\}$ is a stationary process. Under eqn. (5.4.1), the first differences of observations at equally spaced times satisfy the stationary model (5.4.1). Specifically,

$$R_t = Y(t) - Y(t-1) = \{\mu(t) - \mu(t-1)\} + \int_{t-1}^t W(s)\,\mathrm{d}s$$

$$= v(t) + Z_t, \tag{5.4.2}$$

say, where $\{Z_t\}$ is stationary random sequence. Note, however, that the argument leading to eqn. (5.4.2) has implications for the autocovariance structure of $\{Z_t\}$. See Section 3.5 and Sandland and McGilchrist (1979) for elaboration of this point.

Once we are satisfied that the data, possibly after transformation, conform to eqn. (5.1.1), the next stage is to formulate a model for $\gamma(u)$, the autocovariance function of the deviations about the trend. To do this, we subtract from the data our provisional estimates of the underlying trends and examine the sample variogram of the resulting series of residuals; recall from Section 2.5 that the variogram of a stationary process with autocovariance function $\gamma(u)$ is the function $V(u) = \gamma(0) - \gamma(u)$.

Let y_{ij} denote the jth observation on the ith series, t_{ij} the time at which y_{ij} is observed, $\hat{\mu}_{ij}$ our provisional estimate of $\mu_{ij} = E[Y_{ij}]$ and $r_{ij} = y_{ij} - \hat{\mu}_{ij}$ the corresponding residual. Then, the sample variogram $\bar{v}(u)$ is the average value of $\frac{1}{2}(r_{ij} - r_{ik})^2$, averaging being over all the series i, and over all pairs j, k *within* each series such that $|t_{ij} - t_{ik}| = u$. This is an obvious extension of the definition for a single series, as given in Section 2.5. Additionally, we can exploit the mutual independence of the m series to estimate $\gamma(0)$ as the average value of $\frac{1}{2}(r_{ij} - r_{hk})^2$ over all j, k and all i, h with $h \neq i$. Call this estimate $\bar{v}_\infty$.

A word of caution is appropriate at this point. We have already seen in Section 2.6 that subtraction of a moving average estimate of trend induces spurious autocovariance structure into the residual series. Spurious structure can similarly be induced by subtracting an over-smooth parametric estimate of trend.

Example 5.3. Simulated data with a quadratic trend
Figure 5.4a shows a set of simulated data consisting of 20 series of 10 observations at equally spaced times. Each series consists of a quadratic trend, common to all series, with additive white noise deviations about the trend. The curvature in the trend is just noticeable in Fig. 5.4a, but obvious in Fig. 5.4b which shows the 20 sequences of residuals after fitting a *linear* trend by ordinary least squares.

Figure 5.5a shows the sample variogram of these residuals. The structure in this variogram would appear to indicate positive correlation at both short and long lags, where $\bar{v}(u)$ is less than $\bar{v}_\infty$, and negative correlation at intermediate lags, where $\bar{v}(u)$ is greater than $\bar{v}_\infty$. However, this structure is a spurious by-product of the use of an incorrect trend model. Figure 5.5b shows the sample variogram of residuals from a quadratic trend model, again fitted by ordinary least squares. Now $\bar{v}(u) \simeq \bar{v}_\infty$ at all lags u indicating, correctly, uncorrelated random deviations about the trend.

If we estimate the μ_{ij} by averaging all available observations at each time point within each treatment group, these difficulties are greatly diminished. A formal analysis proceeds as follows.

Consider m mutually independent time series, each of length l and observed at a common set of time points. Let Y_{ij} denote the jth observation on the ith series. Write $\mu_j = E[Y_{ij}]$ and $\sigma_{jk} = \text{cov}(Y_{ij}, Y_{ik})$, noting that these quantities do not depend on i. Define $\bar{Y}_j = m^{-1} \sum_{i=1}^{m} Y_{ij}$, and *residuals*

$$R_{ij} = Y_{ij} - \bar{Y}_j$$

$$= m^{-1}\left\{(m-1)Y_{ij} - \sum_{p \neq i} Y_{pj}\right\}$$

$$= m^{-1}\left\{(m-1)Z_{ij} - \sum_{p \neq i} Z_{pj}\right\}, \tag{5.4.3}$$

where $Z_{ij} = Y_{ij} - \mu_{ij}$. We shall show that the R_{ij} have essentially the same second-order properties as the 'true' residuals Z_{ij}.

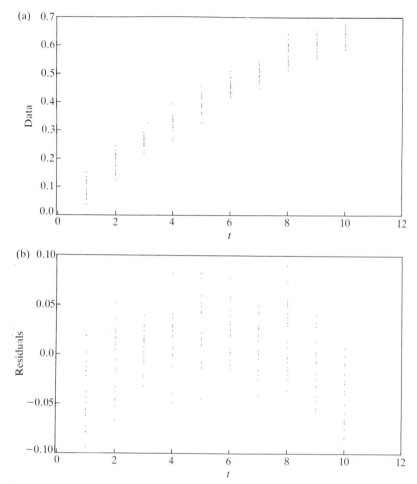

Fig. 5.4. Simulated repeated measurement data with a quadratic trend. (a) Data. (b) Residuals after fitting a *linear* trend by ordinary least squares.

Firstly, note that $E[R_{ij}] = 0$ for all i and j, whatever the form of the underlying μ_{ij}. Secondly, use (5.4.3) to deduce that

$$\text{cov}(R_{ij}, R_{ik}) = m^{-2}E\left[\left\{(m-1)Z_{ij} - \sum_{p \neq i} Z_{pj}\right\}\left\{(m-1)Z_{ik} - \sum_{q \neq i} Z_{qk}\right\}\right]$$

$$= m^{-2}\left\{(m-1)^2 E[Z_{ij}Z_{ik}] - (m-1)\sum_{p \neq i} E[Z_{pj}Z_{ik}]\right.$$

$$\left. - (m-1)\sum_{q \neq i} E[Z_{qk}Z_{ij}] + \sum_{p \neq i}\sum_{q \neq i} E[Z_{pj}Z_{qk}]\right\}. \qquad (5.4.4)$$

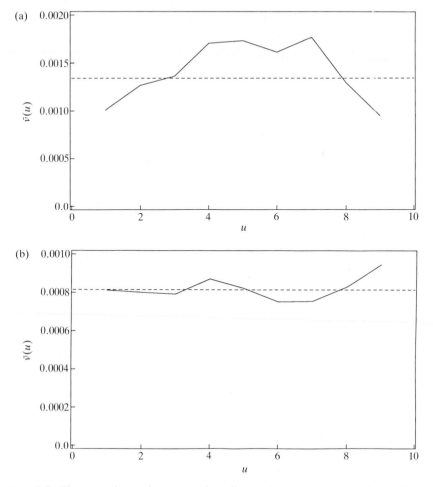

Fig. 5.5. The sample variograms of ordinary least squares residuals from simulated repeated measurements data with a quadratic trend. In each case, the dashed horizontal line represents $\bar{v}_\infty$. (a) After fitting a linear trend. (b) After fitting a quadratic trend.

Because the m series $\{Z_{ij}\}$ are mutually independent, $E[Z_{pj}Z_{qk}] = 0$ unless $p = q$. Also,

$$E[Z_{pj}Z_{pk}] = \mathrm{cov}(Y_{pj}, Y_{pk}) = \sigma_{jk}.$$

Using these results, eqn. (5.4.3) reduces to

$$\mathrm{cov}(R_{ij}, R_{ik}) = m^{-2}\{(m-1)^2\sigma_{jk} + (m-1)\sigma_{jk}\}$$
$$= \{(m-1)/m\}\sigma_{jk}. \tag{5.4.5}$$

In particular, if we denote by $V_z(u)$ and $V_r(u)$ the variograms of the series $\{Z_{ij}\}$ and $\{R_{ij}\}$ respectively, then

$$V_z(u) = \{m/(m-1)\}V_r(u). \tag{5.4.6}$$

Now, consider the covariance between residuals from different series,

$$\text{cov}(R_{ij}, R_{hk}) = m^{-2}E\left[\left\{(m-1)Z_{ij} - \sum_{p\neq i}Z_{pj}\right\}\left\{(m-1)Z_{hk} - \sum_{q\neq h}Z_{qk}\right\}\right]$$

$$= m^{-2}\left\{(m-1)^2 E[Z_{ij}Z_{hk}] - (m-1)\sum_{p\neq i}E[Z_{pj}Z_{hk}]\right.$$

$$\left. - (m-1)\sum_{q\neq h}E[Z_{ij}Z_{qk}] + \sum_{p\neq i}\sum_{q\neq h}E[Z_{pj}Z_{qk}]\right\}.$$

Because $E[Z_{pj}Z_{qk}] = 0$ whenever $p \neq q$, and $E[Z_{pj}Z_{pk}] = \sigma_{jk}$, this expression reduces to

$$\text{cov}(R_{ij}, R_{hk}) = m^{-2}\{-2(m-1)\sigma_{jk} + (m-2)\sigma_{jk}\}$$

$$= -m^{-1}\sigma_{jk}. \tag{5.4.7}$$

The practical implications of these results are as follows. Firstly, eqn. (5.4.6) shows that the sample variogram of the observed residuals r_{ij} is a biased estimator for $V_z(u)$, but can be made unbiased by a simple multiplicative correction. Secondly, from eqns (5.4.5) and (5.4.7) we can deduce that in general, $\bar{v}_\infty$ is also a biased estimator for the process variance, $\gamma(0)$. To see this, combine eqns (5.4.5) and (5.4.7) to give

$$\tfrac{1}{2}E[(R_{ij} - R_{hk})^2] = \text{var}(R_{ij}) - \text{cov}(R_{ij}, R_{hk})$$

$$= \gamma(0)\{1 - m^{-1}(1 - \rho_{jk})\},$$

where ρ_{jk} denotes the autocorrelation between Z_{ij} and Z_{ik}. Thus, the expectation of $\bar{v}_\infty$ is $\gamma(0)\{1 - m^{-1}(1 - \bar{\rho})\}$, where $\bar{\rho}$ denotes an average value of all the relevant ρ_{jk}. Typically all the ρ_{jk}, and therefore $\bar{\rho}$, lie between 0 and 1, suggesting that $\bar{v}_\infty$ has smaller bias than $\bar{v}(u)$.

Strictly, the above analysis breaks down as soon as we relax the requirement of a common set of observation times for all series. However, it is reasonable to suppose that the results will hold approximately if there is a high degree of commonality amongst the m sets of obervation times. Without this, estimation of the underlying trends by simple averaging at each time point is not a practical proposition. Note also that when there are two or more treatment groups, the value of m in the formulae (5.4.5)–(5.4.7) refers to the number of replicate series *within* each treatment group, not to the number of series in the complete set of data.

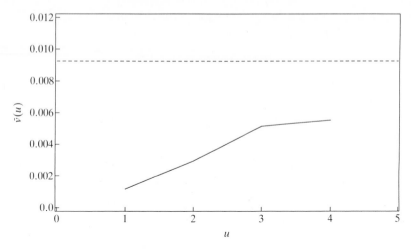

Fig. 5.6. The sample variogram of log-transformed body weights of rats, after subtracting separate means for each time-point in each treatment group. The dashed horizontal line represents $\bar{v}_\infty$.

Example 5.4. Sample variogram of log-transformed body weights of rats
Figure 5.6 shows the sample variogram of the data on log-transformed body weights of rats, obtained as suggested above by first subtracting from each log-weight the mean of all observations from the appropriate time point and treatment group. Because the number of animals per treatment group is either 7 or 10, we have not attempted to remove the bias in the sample variogram, supposing this to be small.

With only four values of $\bar{v}(u)$ available, we should beware of attempting an over-precise interpretation of the functional form of the underlying population variogram $V(u)$. Nevertheless, three distinct features are clear: the pattern of $\bar{v}(u)$ increasing with u but levelling out by $u = 4$ is consistent with a positive short-range autocorrelation decaying rapidly to zero; a rough extrapolation of $\bar{v}(u)$ back to zero suggests that $V(0) \approx 0$, i.e. that measurement error is small; the substantial difference between $\bar{v}_\infty$ and $\bar{v}(4)$ indicates substantial between-animal variation. Finally, the functional form of $\bar{v}(u)$ appears to be consistent with the Gaussian correlation function (5.3.2),

$$g(u) = \exp(-\alpha u^2),$$

corresponding to a theoretical variogram

$$V(u) = \tau^2 + \sigma^2\{1 - \exp(-\alpha u^2)\} \tag{5.4.8}$$

and variance

$$V_\infty = \tau^2 + v^2 + \sigma^2. \tag{5.4.9}$$

Within this parametric model, we can use Fig. 5.6 to make rough estimates of the parameters, by substituting a few values from the *sample* variogram for the theoretical quantities $V(u)$ and V_∞ in eqns (5.4.8) and (5.4.9), and solving the resulting set of simultaneous equations.

Substituting $\bar{v}_\infty$ for V_∞ in eqn. (5.4.9) gives

$$\tau^2 + v^2 + \sigma^2 \approx 0.0092. \qquad (5.4.10)$$

A rough estimate of the asymptote of $\bar{v}(u)$ at large u is 0.006, suggesting

$$\tau^2 + \sigma^2 \approx 0.006. \qquad (5.4.11)$$

Combining eqns (5.4.10) and (5.4.11) we deduce that

$$v^2 \approx 0.0032.$$

Equally rough back-extrapolation of $\bar{v}(u)$ to $u = 0$ gives

$$\tau^2 \approx 0.0005$$

and hence

$$\sigma^2 \approx 0.0055.$$

Finally, substitution of $\bar{v}(1)$ into eqn. (5.4.8) gives

$$0.0015 \approx 0.0005 + 0.0055(1 - e^{-\alpha}),$$

or

$$\alpha \approx 0.20.$$

We do not suggest that this rough-and-ready procedure provides sensible final estimates of the parameters. Rather, these estimates should be seen as initial guesses, to be refined by the more formal procedures to be discussed in Section 5.5.

Example 5.5. Sample variogram of data on growth of colonies of *paramecium aurelium*

The data from Example 1.5 consist of three series of counts of *paramecium aurelium* in 10% samples drawn at daily intervals from each of three colonies growing in a nutrient medium. For data of this kind, we would expect the variability to increase with the mean: a plausible model for the sampling variability in the data is that, conditional on the colony size N at any given time, the observed count X is binomially distributed with probability distribution

$$P\{X = x \mid N\} = \binom{N}{x} p^x (1-p)^{N-x},$$

where $p = 0.1$. If μ and σ^2 denote the population mean and variance of N, standard manipulations lead to the following results:

$$E(X) = \mu p$$
$$\text{var}(X) = \mu p(1-p) + \sigma^2 p^2 = \mu p + p^2(\sigma^2 - \mu).$$

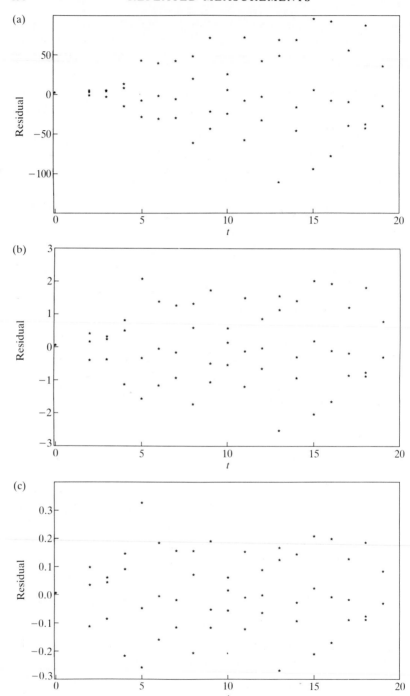

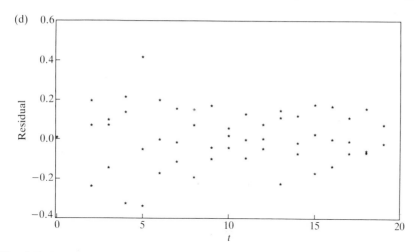

Fig. 5.7. Residuals of transformed counts of *paramecium aurelium*, after subtracting the mean of the three replicates at each time point. (a) No transformation, $y_t = \text{count}$. (b) Square root, $y_t = (\text{count})^{0.5}$. (c) Fourth root, $y_t = (\text{count})^{0.25}$. (d) Logarithm, $y_t = \log(\text{count})$.

If the colony size N at each time point follows a Poisson distribution then $\sigma^2 = \mu$ and $\text{var}(X) = E(X)$: in fact, in this case Z also follows a Poisson distribution, and a square-root transformation would lead to data with variance approximately constant over time.

A somewhat more empirical approach to the choice of a transformation to stabilize the variance of the data over time is to compute residuals by subtracting the mean of the three values at each time point, and to plot the residuals against time. Figure 5.7 shows the result for the untransformed counts and for square-root, fourth root and log-transformed counts. Figure 5.7a shows variability clearly increasing over time, whilst Fig. 5.7d suggests the opposite. Figure 5.7c suggests an approximately constant variance, and we therefore use a fourth root transformation, $y_t = (\text{count})^{0.25}$, for the subsequent analysis.

Figure 5.8 shows the resulting sample variogram. In this case, $m = 3$ and we have multipled each $\bar{v}(u)$ by $m/(m-1) = 1.5$ to remove the bias. We have also multiplied $\bar{v}_\infty$ by 1.5 although, as noted above, this is likely to over-correct for the bias.

The sample variogram $\bar{v}(u)$ shows a gentle increase between lags 1 and 11, but thereafter decreases quite sharply. With only three series of eighteen observations available, the values of $\bar{v}(u)$ at large lags u are averages of small numbers of squared differences and are statistically unreliable. A simple parametric model which is consistent with the

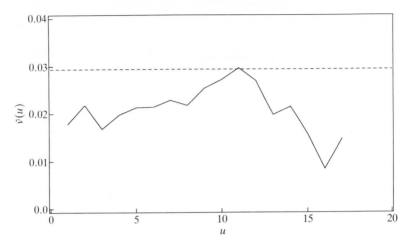

Fig. 5.8. The sample variogram of fourth-root-transformed counts of *paramecium aurelium*, after subtracting the mean of the three replicates at each time point. Each $\bar{v}(u)$ has been multiplied by 1.5 to remove bias. The dashed horizontal line represents the value of $\bar{v}_\infty$, again multiplied by 1.5.

behaviour of the sample variogram at small lags takes

$$\rho(u) = \begin{cases} 1 - \alpha u & : \quad u \le \alpha^{-1} \\ 0 & : \quad u > \alpha^{-1} \end{cases}$$

in (5.3.3), corresponding to a theoretical variogram

$$V(u) = \begin{cases} \tau^2 + \sigma^2 \alpha u & : \quad u \le \alpha^{-1} \\ \tau^2 + \sigma^2 & : \quad u > \alpha^{-1} \end{cases} \qquad (5.4.12)$$

with

$$V_\infty = \tau^2 + v^2 + \sigma^2 \qquad (5.4.13)$$

as before.

As in Example 5.4, we can calculate rough estimates of the parameters from the plot of the sample variogram. This gives $\tau^2 \simeq 0.016$, $v^2 \simeq 0.001$, $\sigma^2 \simeq 0.013$ and $\alpha \simeq 0.1$. Again, these are intended only as initial values for a formal estimation procedure. The value of v^2 close to zero presumably reflects the biological homogeneity of the experimental material.

5.5　Fitting the model: maximum likelihood estimation and testing

We now suppose that we have arrived at a parametric model for the data in question, within the general framework described in Section 5.3 but without the restriction to a common set of observation times. Thus, we assume that for each of $i = 1, \ldots, m$, mean values for the random vector $\mathbf{Y}_i$ of l_i observations on the ith unit are expressible as

$$\boldsymbol{\mu}_i = X_i\boldsymbol{\theta},$$

where $\boldsymbol{\theta}$ is a p-element vector of unknown parameters and X_i a known l_i by p matrix. We also assume that the variance matrix, Σ_i, is of the form implied by eqn. (5.3.3) and write $\Sigma_i = \sigma^2 V_i(\boldsymbol{\phi})$, using the vector $\boldsymbol{\phi}$ to embrace $\phi_1 = v^2/\sigma^2$, $\phi_2 = \tau^2/\sigma^2$ and any additional parameters which define the autocorrelation function $\rho(u)$ in eqn. (5.3.3). For example, if we were to adopt the model (5.3.2) we would put $\phi_3 = \alpha$.

Now, assuming a multivariate Normal distribution for each $\mathbf{Y}_i$, we obtain the log-likelihood for the parameters $\boldsymbol{\theta}$, σ^2 and $\boldsymbol{\phi}$ as

$$L(\boldsymbol{\theta}, \sigma^2, \boldsymbol{\phi}) = -\tfrac{1}{2}\left[n \log \sigma^2 + \sum_{i=1}^{m} \log |\Sigma_i(\boldsymbol{\phi})| \right.$$

$$\left. + \sum_{i=1}^{m} (y_i - X_i\boldsymbol{\theta})'\{V_i(\boldsymbol{\phi})\}^{-1}(y_i - X_i\boldsymbol{\theta})/\sigma^2 \right], \quad (5.5.1)$$

where $n = \sum_{i=1}^{m} l_i$. Given values for the elements of $\boldsymbol{\phi}$, we can write the maximum likelihood estimators for $\boldsymbol{\theta}$ and σ^2 explicitly as

$$\hat{\boldsymbol{\theta}}(\boldsymbol{\phi}) = \left[\sum_{i=1}^{m} X_i'\{V_i(\boldsymbol{\phi})\}X_i \right]^{-1}\left[\sum_{i=1}^{m} X_i'\{V_i(\boldsymbol{\phi})\}^{-1}y_i \right] \quad (5.5.2)$$

and

$$\hat{\sigma}^2(\boldsymbol{\phi}) = n^{-1} \sum_{i=1}^{m} \{\mathbf{y}_i - X_i\hat{\boldsymbol{\theta}}(\boldsymbol{\phi})\}'\{V_i(\boldsymbol{\phi})\}^{-1}\{\mathbf{y}_i - X_i\hat{\boldsymbol{\theta}}(\boldsymbol{\phi})\}. \quad (5.5.3)$$

By substituting expressions (5.5.2) and (5.5.3) back into eqn. (5.5.1) we obtain a reduced form of the log-likelihood,

$$L(\boldsymbol{\phi}) = -\tfrac{1}{2}\left[n \log\{\sigma^2(\boldsymbol{\phi})\} + \sum_{i=1}^{m} \log\{|V_i(\boldsymbol{\phi})|\} \right], \quad (5.5.4)$$

which involves only $\boldsymbol{\phi}$. Except in very special circumstances, we have to use numerical methods to maximize (5.5.4). Suitable general-purpose algorithms for non-linear optimization are now widely available, for example in the NAG and IMSL subroutine libraries. The analyses reported in the remainder of this chapter use the simplex algorithm of Nelder and Mead (1965), which is unsophisticated but robust and easy to use.

After maximizing (5.5.4) to obtain $\hat{\boldsymbol{\phi}}$, the maximum likelihood estimates of $\boldsymbol{\theta}$ and σ^2 are $\hat{\boldsymbol{\theta}} = \hat{\boldsymbol{\theta}}(\hat{\boldsymbol{\phi}})$ and $\hat{\sigma}^2 = \hat{\sigma}^2(\hat{\boldsymbol{\phi}})$, obtained by substitution of $\hat{\boldsymbol{\phi}}$ into eqns (5.5.2) and (5.5.3), respectively.

With regard to standard errors for the elements of $\hat{\boldsymbol{\theta}}$, we know that for any fixed σ^2 and $\boldsymbol{\phi}$, the variance matrix of $\hat{\boldsymbol{\theta}}(\boldsymbol{\phi})$ is

$$\text{var}\{\hat{\theta}(\boldsymbol{\phi})\} = \sigma^2 \left[\sum_{i=1}^{m} X_i' \{V_i(\boldsymbol{\phi})\}^{-1} X_i \right]^{-1}.$$

It follows that the approximate variance matrix of $\hat{\boldsymbol{\theta}}$ is

$$\text{var}(\hat{\boldsymbol{\theta}}) \simeq \hat{\sigma}^2 \left[\sum_{i=1}^{m} X_i' \{V_i(\hat{\boldsymbol{\phi}})\}^{-1} X_i \right]^{-1}. \tag{5.5.5}$$

A convenient way to test hypotheses about $\boldsymbol{\theta}$ is to compare the maximized log-likelihoods under the various hypotheses of interest. If L_0 and L_1 represent the maximized log-likelihoods for null and alternative models with q and $p > q$ parameters θ_i, respectively, the generalized likelihood ratio statistic to test the adequacy of the null model is $D = 2(L_1 - L_0)$, and the distribution of D under the null model is chi-squared on $p - q$ degrees of freedom (cf. Section 4.8).

Example 5.6. Analysis of data on log-transformed body-weights of rats
We are now in a position to complete our analysis of the data from Example 1.6. Figure 5.9 shows the observed mean response profiles for the log-transformed data. The response in each treatment group is approximately quadratic in time, and we therefore assume that the mean response in treatment group i at time t is of the form

$$\mu_i(t) = a_i + b_i t + c_i t^2. \tag{5.5.6}$$

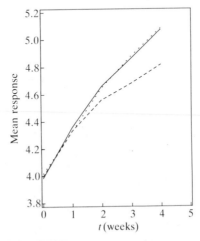

Fig. 5.9. Mean response profiles for log-transformed body weights of rats. —, Control; - - - - -, thyroxin; – – –, thiouracil.

In Example 5.4, we identified a parametric model for the covariance structure of the data, and used the sample variogram to suggest approximate parameter values. We can now use these as initial values for numerical maximization of the likelihood function. The resulting maximum value of the log-likelihood is 383.3.

The essential question for these data is whether the three mean response profiles in eqn. (5.5.6) are different. One way to answer this question is to maximize the likelihood but with the nine parameters a_i, b_i, and c_i replaced by three parameters a, b, and c to define a common mean response profile in all three treatment groups. The resulting maximized log-likelihood is 362.4, the generalized likelihood ratio statistic is $D = 41.8$ on six degrees of freedom, and we conclude that there are highly significant differences amongst the $\mu_i(t)$, the attained significance level being $p < 0.001$. Inspection of Fig. 5.9 suggests strongly that thiouracil inhibits growth whereas the effect of thyroxin is indistinguishable from that of the control. To confirm this, we maximize the log-likelihood for a model with six mean value parameters obtained by setting $a_1 = a_2$, $b_1 = b_2$ and $c_1 = c_2$. The maximized value is 382.4. Comparing this with the maximized value for the full model gives a generalized likelihood ratio statistic of $D = 1.8$ on three degreees of freedom. This is clearly non-significant ($p \approx 0.61$), confirming that the six-parameter model is adequate. The maximum likelihood estimates, and approximate standard errors of the mean value parameter estimates, are given in Table 5.1.

To complete the analysis of these data, Fig. 5.10 compares the observed and fitted mean response profiles and variograms. In both cases there is good agreement, suggesting that the model is a reasonable representation of the data.

Example 5.7. Analyis of data on growth of colonies of *paramecium aurelium*
In similar vein, we can complete an analysis of the data discussed in Examples 1.6 and 5.5, fitting a parametric model by the method of maximum likelihood.

Figure 5.11 shows the fourth-root transformed data. The simplest *linear* model for the mean response profile which gives a reasonable fit to the data is a quartic polynomial,

$$\mu(t) = a + bt + ct^2 + dt^3 + et^4.$$

Figure 5.12 summarizes the fit to the data, in terms of the observed and fitted mean response profiles in Fig. 5.12a, and variograms in Fig. 5.12b.

Although the quartic polynomial gives a reasonable fit to the data within the observed time-span, it must be emphasized that it would be

Table 5.1. Maximum likelihood estimates for final model fitted to logarithms of body weights of rats

(a) Mean value parameters: $\mu_i(t) = a_i + b_i t + c_i t^2$

Parameter	Estimate	Standard error
$a_1 = a_2$	4.00	0.023
$b_1 = b_2$	0.366	0.016
$c_1 = c_2$	−0.024	0.004
a_3	3.99	0.030
b_3	0.343	0.020
c_3	−0.033	0.005

(b) Covariance parameters

Parameter	Estimate
σ^2	0.00483
$\phi_1 = \tau^2/\sigma^2$	0.105
$\phi_2 = \nu^2/\sigma^2$	0.761
$\phi_3 = \alpha$	0.253

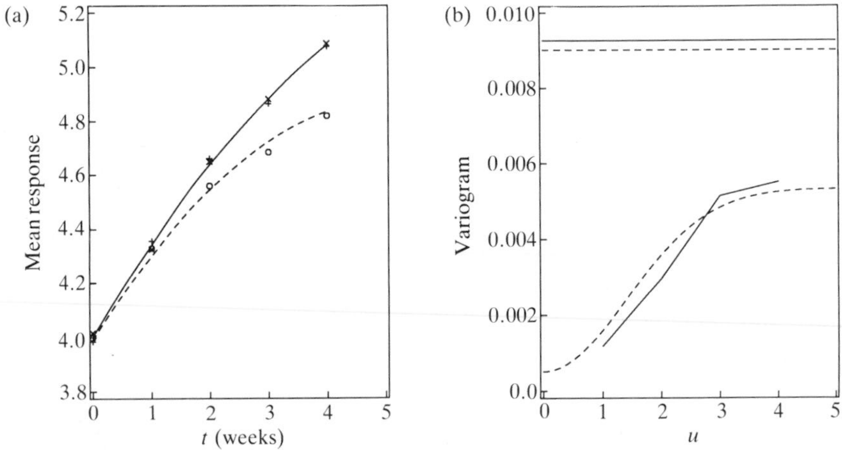

Fig. 5.10. Observed and fitted mean response profiles and variograms for log-transformed body weights of rats. (a) Mean response profiles: +, observed control; ×, observed thyroxin; ○, observed thiouracil; —, fitted control and thyroxin; ---, fitted thiouracil. (b) Variograms, with horizontal lines to represent process variance; —, observed; ---, fitted.

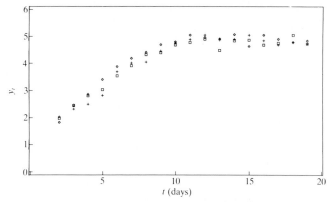

Fig. 5.11. Fourth-root-transformed counts of *paramecium aurelium*. +, first replicate; □, second replicate; ◇, third replicate.

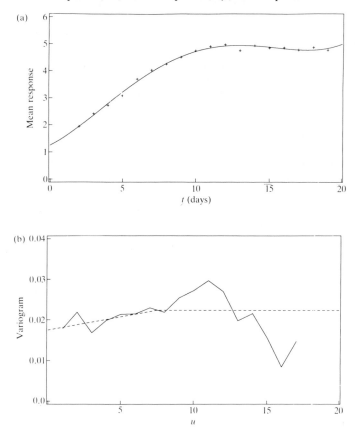

Fig. 5.12. Observed and fitted mean response profiles and variograms for fourth-root-transformed counts of *paramecium aurelium*. (a) Mean response profiles: +, observed; —, fitted. (b) Variograms: —, observed; – – –, fitted.

disastrous for extrapolation. For example, beyond $t = 20$ the fitted mean response increases rapidly, reaching a value of 21.2 by $t = 30$. A *non-linear* model which approaches a limiting value as t increases would seem to be preferable on scientific grounds. Note that in contrast to the previous example the form of the mean response $\mu(t)$ is of direct interest, rather than of indirect interest as an expression of treatment effects.

There is an extensive literature on non-linear modelling of growth data. See, for example, Ratkowsky (1983). Almost all of this literature assumes that successive observations exhibit uncorrelated random variation about the underlying mean response profile. One exception is Glasbey (1979), whose consideration of autocorrelated random variation is similar in spirit to the approach taken here for *linear* mean response models.

5.6 Case-study: analysis of data on protein content of milk samples

We recall from Example 1.7 that these data record the protein content of milk samples taken at weekly intervals for up to 19 weeks from a total of 79 cows in three treatment groups, the treatments corresponding to the three diets; (1) barley, (2) mixed (barley and lupins), and (3) lupins. The three observed mean response profiles displayed in Fig. 1.9 all showed a sharp drop over the first three weeks of the experiment, followed by a more or less constant mean response over the remaining 16 weeks.

We first calculate the empirical variogram of the residuals after subtracting a separate estimated mean value for each of the 57 treatment by time combinations. Since the number of animals per treatment group is at least 25, the bias in the sample variogram is negligible, and we made no adjustment for it. Figure 5.13 shows the result. If we discount the sudden increase in $\bar{v}(u)$ at $u = 18$, where there is least information and therefore largest sampling variation, the empirical variogram is consistent with the model postulated in Section 5.3 incorporating an exponential correlation function,

$$\rho(u) = \exp(-\alpha u),$$

and $\alpha \simeq 0.2$. Furthermore, the value of $\bar{v}(\infty)$ and the apparent asymptote of $\bar{v}(u)$ together suggest that $v^2 \simeq 0.01$, whilst crude extrapolation of $\bar{v}(u)$ to $u = 0$ suggests that $\tau^2 \simeq 0.02$.

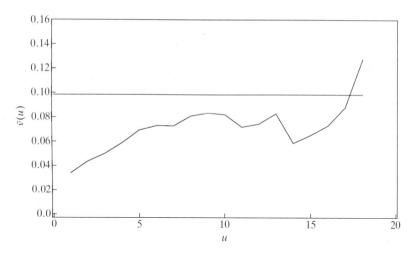

Fig. 5.13. The sample variogram of protein content of milk samples, after subtracting separate means for each time point in each treatment group. The horizontal lines represents $\bar{v}_\infty$.

With regard to the mean response profiles, the practical context suggests that the variation in the mean response over time might be seasonal. However, it is clearly not sinusoidal. A simple model which appears to fit the data well is a superposition of two sinusoids to describe the mean response profile within each group, with parallel profiles for the three groups. In algebraic terms we express the mean response at time t in treatment group i as

$$\mu_i(t) = \mu_i + \beta_1 \cos(\omega t) + \gamma_1 \sin(\omega t) + \beta_2 \cos(2\omega t) + \gamma_2 \sin(2\omega t), \quad (5.6.1)$$

where $\omega = \pi/26$ to correspond to a periodicity of 52 weeks.

The results of fitting this model to the data by maximum likelihood are summarized in Table 5.2. The generalized likelihood ratio statistic for treatment differences, i.e. unequal μ_i, is 15.22 on two degrees of freedom ($p \approx 0.0005$). The conclusion is that diet affects percentage protein, with barley giving the highest percentage and lupins the lowest.

The goodness-of-fit of the model is summarized graphically in Fig. 5.14. Figure 5.14a suggests a good fit to the assumed covariance structure, again discounting the large value of $\bar{v}(18)$. Figure 5.14b

Table 5.2. Maximum likelihood estimates for seasonal model fitted to data
on protein content of milk samples

(a) Mean value parameter:
$\mu_i(t) = \mu_i + \alpha_1 \cos(\omega t)$
$+ \beta_1 \sin(\omega t) + \alpha_2 \cos(2\omega t)$
$+ \beta_2 \sin(2\omega t)$, $w = \pi/26$

Parameter	Estimate
μ_1	5.82
μ_2	5.72
μ_3	5.61
α_1	−0.60
β_1	−3.34
α_2	−1.04
β_2	0.39

(b) Covariance parameters

Parameter	Estimate
σ^2	0.071
$\phi_1 = \tau^2/\sigma^2$	0.305
$\phi_2 = v^2/\sigma^2$	0.065
$\phi_3 = \alpha$	0.184

suggests a tolerably good fit to the assumed mean response profiles,
except possibly towards the end of the experiment.

Closer examination of the fitted mean response profiles, however,
brings to light a disturbing feature. This is exhibited graphically in Fig.
5.15, which plots the fitted response on the first diet over a two-year
period. The seasonal variation is enormous, and quite unrealistic.

An alternative view of the data is to regard the first three weeks as a
'settling-in' period, during which mean percentage protein apparently
decreases linearly. This suggests an alternative model for the mean
response profiles $\mu_i(t)$, namely

$$\mu_i(t) = \begin{cases} \mu_i + \beta t & : \quad t \leq 3 \\ \mu_i + 3\beta & : \quad t > 3. \end{cases} \qquad (5.6.2)$$

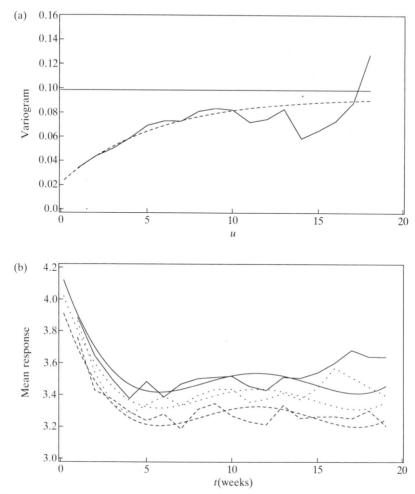

Fig. 5.14. Observed and fitted variograms and mean response profiles for protein content of milk samples. (a) Variograms: —, observed; – – –, fitted. (b) Mean response profiles, with observed shown as connected line segments and fitted as smooth curves: —, barley; - - -, barley and lupins; – – –, lupins.

This model is still linear in the sense of Section 5.5, turns out to fit the data as well as the seasonal model (5.6.1), and gives essentially the same conclusion with regard to treatment effects. Figure 5.16 summarizes the fit of (5.6.2) to the observed mean response profiles. Table 5.3 gives the maximum likelihood estimates of the model parameters. Note in particular that the estimated differences, $\hat{\mu}_i - \hat{\mu}_j$, and the estimates of the parameters defining the covariance structure of the model, are similar to

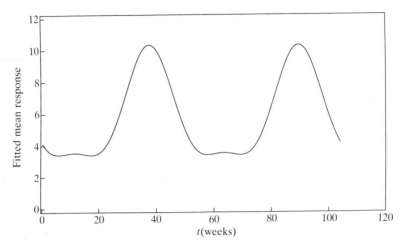

Fig. 5.15. Extrapolation of the fitted mean response profile for protein content of milk samples from cows fed a diet of barley.

those obtained using the seasonal model. Furthermore, the standard error of each estimated treatment difference, $\hat{\mu}_i - \hat{\mu}_j$, is approximately 0.05 under either model.

This example illustrates very well the general point that models which fit the data equally well may differ enormously in their wider interpretation. Common sense suggests, in retrospect, that the 'settling-in' model

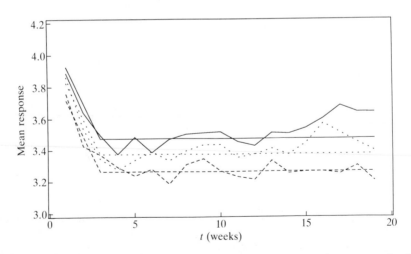

Fig. 5.16. Observed and fitted mean response profiles for protein content of milk samples, assuming a three-week settling-in period. —, barley; - - - -, barley and lupins; – – –, lupins.

Table 5.3. Maximum likelihood estimates for 'settling-in' model fitted to data on protein content of milk samples

(a) Mean value parameter:

$$\mu_i(t) = \begin{cases} \mu_i + \beta t & : \quad t \leqslant 3 \\ \mu_i + 3\beta & : \quad t > 3 \end{cases}$$

Parameter	Estimate
μ_1	4.15
μ_2	4.05
μ_3	3.94
β	−0.225

(b) Covariance parameters

Parameter	Estimate
σ^2	0.068
$\phi_1 = \tau^2/\sigma^2$	0.338
$\phi_2 = v^2/\sigma^2$	0.063
$\phi_3 = \alpha$	0.172

(5.6.2) is more sensible than the seasonal model (5.6.1), but this conclusion requires a piece of non-statistical reasoning to add to the statistical evidence.

5.7 Further reading

In this chapter, we have followed Diggle (1988) in using the method of maximum likelihood to estimate jointly the parameters which define the mean response and those which define the covariance structure. Restricted maximum likelihood estimation (Thompson 1962; Cullis and McGilchrist 1988), also known as marginal likelihood estimation (Tunnicliffe–Wilson 1989), is a variant of maximum likelihood which effectively estimates these two sets of parameters separately and leads to reduced bias in the estimation of the covariance parameters. The improvement over maximum likelihood can be substantial when the model for the mean response contains many parameters.

We have considered only the case of Normally distributed repeated measurements. Alternative methodologies for discrete or categorical repeated observations include those described in Koch *et al.* (1977), Stiratelli *et al.* (1984), and Zeger and Liang (1986).

Koch *et al.* (1980) and Diggle and Donnelly (1989) give selected bibliographies on the analysis of repeated measurements.

6
Fitting autoregressive integrated moving average processes to data

6.1 Introduction: ARIMA processes as models for non-stationary time series

In Section 3.4 we introduced the class of autoregressive moving average (ARMA) processes and discussed the second-order properties of these processes. We also pointed out that the value of ARMA processes as models for time-series data lies primarily in their ability to approximate a wide range of second-order behaviour using only a small number of parameters. For this reason, methodology for fitting ARMA-processes to time-series data is of interest in its own right. However, an additional motivation for considering this methodology is its central role in the Box–Jenkins approach to forecasting, which we shall describe in Chapter 7.

One extension to the ARMA class of processes which greatly enhances their value as empirical descriptors of *non-stationary* time series is the class of *autoregressive integrated moving average*, or ARIMA, processes. We say that $\{Y_t\}$ is an ARIMA process of order p, d, q, written $Y_t \sim \text{ARIMA}(p, d, q)$, if the dth difference of Y_t is a stationary, invertible ARMA process of order p, q. Thus, using the back-shift operator notation,

$$\phi(B)(1 - B)^d Y_t = \theta(B)Z_t, \qquad (6.1.1)$$

where $\{Z_t\}$ is white noise, $\phi(\cdot)$ and $\theta(\cdot)$ are polynomials of degree p and q respectively with all roots of the polynomial equations $\phi(z) = 0$ and $\theta(z) = 0$ outside the unit circle.

In Example 1.8, we gave a simple example of an ARIMA process, namely the random walk,

$$(1 - B)Y_t = Z_t,$$

or $Y_t \sim \text{ARIMA}(0, 1, 0)$. Figure 6.1 shows a realization of an ARIMA $(1, 1, 0)$ process,

$$(1 - 0.7B)(1 - B)Y_t = Z_t.$$

This series exhibits a characteristic feature of many ARIMA processes,

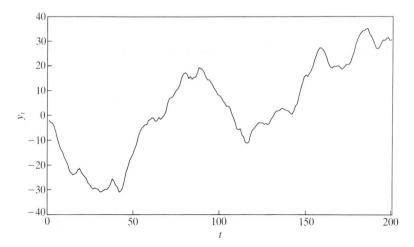

Fig. 6.1. A simulated realization of an ARIMA$(1, 1, 0)$ process, $(1 - 0.7B)(1 - B)Y_t = Z_t$, $n = 200$ observations.

namely a tendency for relatively long stretches of the data to follow an apparent trend which itself evolves in a random fashion. For example, the series appears to follow a non-linear rising trend between times 50 and 80; an approximately linear falling trend tetween times 90 and 110; and so on. To see why this happens, note that the ARIMA process Y_t can be represented as

$$Y_t = Y_{t-1} + W_t,$$

where $\{W_t\}$ is a zero-mean, stationary AR(1) process with autoregressive parameter 0.7. This means that $E(Y_t) = E(Y_{t-1})$, i.e. the average value of Y_t does not depend on t and there is no genuine trend. However, because the process $\{W_t\}$ is strongly autocorrelated, it tends to take relatively long excursions above or below zero, leading to a corresponding increasing or decreasing stretch of Y_t values, respectively. We use the term 'stochastic trend' to describe this phenomenon. Qualitatively, it is reminiscent of the behaviour of many economic time series, such as the log-wool-price-ratio series of Example 1.3.

ARIMA processes provide a theoretical justification for differencing a non-stationary series in order to achieve stationarity. However, this justification is in terms of a very specialized form of non-stationarity. For cautionary remarks on the use of differencing as a general approach to trend removal, see Section 2.3.

Fitting an ARIMA process to an observed time series proceeds in three stages: *identification* of the order (p, d, q); *estimation* of the model parameters; *diagnostic checking* of the fitted model. In the next three sections we consider each of these stages in turn.

6.2 Identification

The first stage in the identification of an ARIMA process is to examine a time plot of the data, $\{y_t\}$, for evidence of non-stationarity. If the data appear to be stationary, no differencing is called for, and we have provisionally identified $d = 0$ in eqn. (6.1.1). Note in particular our use of the qualifying adverb 'provisionally'. Identification of appropriate values for p, d, and q involves subjective judgement and it is good sense to keep open the possibility of changing the model if, at a later stage, the provisional identification is found wanting.

If the data appear to be non-stationary, we successively difference the series until its time plot appears to be stationary. In practice, $d = 1$ or 2 often suffices. When the time plot is ambiguous, it may be useful to compute the correlogram of the data, as a correlogram which fails to decay to zero is indicative of non-stationarity.

Recall from Section 2.5 that the correlogram of a time series $\{y_t : t = 1, \ldots, n\}$ is a graph of the sample autocorrelation coefficients r_k against the corresponding lags k, each r_k being defined as

$$r_k = g_k / g_0,$$

where

$$g_k = \sum_{t=k+1}^{n} (y_t - \bar{y})(y_{t-k} - \bar{y})/n.$$

For an underlying white noise process $\{Y_t\}$, and for large n, the approximate sampling distribution of each r_k is Normal, with mean zero and variance $1/n$, a result previously given as eqn. (2.5.3). This suggests using the limits $\pm 2/\sqrt{n}$ to assess individual r_k for significant departure from zero. These limits are useful as a rough guide to interpreting a correlogram, but should not be interpreted rigidly. Note in particular that successive r_k are themselves correlated.

In general, the significance of individual r_k is less important than their overall pattern. Specifically, we search for a pattern in the r_k which can be modelled adequately by an ARMA process. An obvious difficulty in this is that any underlying pattern in the theoretical autocorrelation coefficients will be obscured by sampling fluctuations in the r_k.

The simplest type of pattern to detect from visual inspection of the correlogram is a *cut-off*. By this, we mean that all r_k for k greater than some integer q are approximately zero. According to the results in Section 3.4.2, this indicates an MA(q) process as a possible model for the data.

Smoothly decaying autocorrelations are more difficult to detect by visual inspection, except perhaps when the decay takes a particularly simple form such as the exponential decay associated with an AR(1) process. For this reason, we introduce a variant of the correlogram, the *partial correlogram,* which has a cut-off for an underlying autoregressive process.

To construct the partial correlogram, we successively fit autoregressive processes of order $1, 2, \ldots$, and, at each stage, define the partial autocorrelation coefficient, a_k, to be the estimate of the final autoregressive coefficient: thus, a_k is the estimate of α_k in an $AR(k)$ process. Clearly, if the underlying process is $AR(p)$, then $\alpha_k = 0$ for all $k > p$, and the partial correlogram should show a cut-off after lag p.

The simplest way to construct the partial correlogram is via the sample analogues of the Yule–Walker equations (3.4.15). These equations express the relationship between the autoregressive parameters α_k and the theoretical autocorrelations ρ_k as

$$\rho_k = \sum_{l=1}^{p} \alpha_l \rho_{k-l} \quad : \quad k = 1, \ldots, p,$$

where p is the order of the underlying autoregression. The sample analogue of the above set of equations is

$$r_k = \sum_{l=1}^{p} \hat{\alpha}_{l,p} r_{k-l} \quad : \quad k = 1, 2, \ldots, p, \tag{6.2.1}$$

where we write $\hat{\alpha}_{l,p}$ to emphasize that we are estimating the autoregressive coefficients $\alpha_l : l = 1, \ldots, p$ on the assumption that the underlying process is autoregressive of order p.

Equations (6.2.1) can be solved explicitly by writing them in matrix notation. We write $\mathbf{r} = (r_1, \ldots, r_p)'$, $\hat{\mathbf{\alpha}} = (\hat{\alpha}_{1,p}, \ldots, \hat{\alpha}_{p,p})'$ and R for the p by p matrix with (i, j)th element r_{i-j}. Recall that, by definition, $r_{-k} = r_k$. Then, eqn. (6.2.1) can be written as

$$\mathbf{r} = R\hat{\mathbf{\alpha}}.$$

Because R is an invertible matrix, this gives

$$\hat{\mathbf{\alpha}} = R^{-1}\mathbf{r},$$

and the pth partial autocorrelation coefficient is $a_p = \hat{\alpha}_{p,p}$.

Evaluating the a_p in this way is computationally wasteful if p is large. An alternative, recursive method of calculation due to Durbin (1960) uses the following formulae:

$$\hat{\alpha}_{j,p+1} = \hat{\alpha}_{j,p} - \hat{\alpha}_{p+1,p+1}\hat{\alpha}_{p-j+1,p} \quad : \quad j = 1, \ldots, p,$$

$$\hat{\alpha}_{p+1,p+1} = \left(r_{p+1} - \sum_{j=1}^{p} \hat{\alpha}_{jp} r_{p+1-j} \right) \Big/ \left(1 - \sum_{j=1}^{p} \hat{\alpha}_{j,p} r_j \right).$$

Interpretation of the partial correlogram proceeds along the same lines as for the correlogram, using $\pm 2/\sqrt{n}$ limits to assess cut-off. The justification for this is a result due to Quenouille (1949), who showed that for an underlying AR(p) process, the approximate sampling distribution of each a_k with $k > p$ is Normal, $a_k \sim N(0, 1/n)$.

To summarize: we now have a means of assessing whether a moving average process ($p = 0$) or an autoregressive process ($q = 0$) is appropriate by looking for evidence of cut-off in the correlogram or partial correlogram respectively. If neither cuts off at a sufficiently small lag, say 2 or 3, we may do better to consider a 'mixed' process, with neither p nor q equal to zero. In this case parsimony suggests provisionally identifying $p = q = 1$.

The complete identification process is summarized in Fig. 6.2.

In conclusion we re-emphasize that model identification as here presented is a subjective process, and that parsimony should be the guiding principle. If in doubt, we should opt for fewer rather than more parameters in the identified model, but include in the third, diagnostic checking, stage an assessment of whether additional parameters are needed to give an adequate fit.

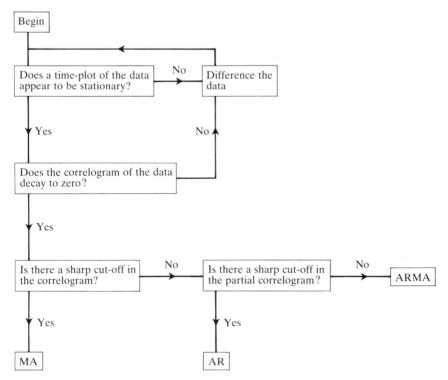

Fig. 6.2. Flowchart illustrating the identification process.

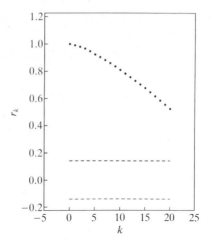

Fig. 6.3. The correlogram of the simulated data from Fig. 6.1. The dashed horizontal lines correspond to the limits $\pm 2/\sqrt{n}$.

Example 6.1. Model identification for simulated data
We illustrate the model identification procedure using the simulated data from Fig. 6.1. Visual inspection of Fig. 6.1 strongly suggests non-stationarity. Were we in any doubt of this, we could compute the correlogram, which is here shown as Fig. 6.3. The very slow decay from $r_0 = 1$ to $r_{20} \simeq 0.5$ reinforces the conclusion that the data are non-stationary. We therefore difference the data, to obtain the series plotted in Fig. 6.4. This does appear to be stationary, as is confirmed by the

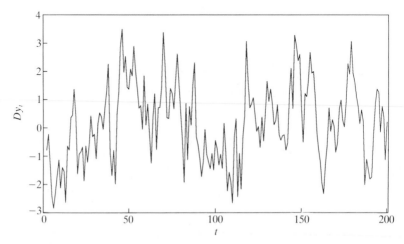

Fig. 6.4. The first difference of the simulated data from Fig. 6.1.

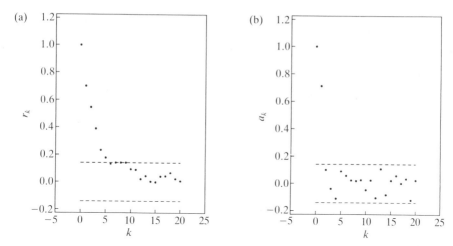

Fig. 6.5. The correlogram and partial correlogram of the first difference of the data from Fig. 6.1. The dashed horizontal lines correspond to the limits $\pm 2/\sqrt{n}$. (a) Correlogram. (b) Partial correlogram.

correlogram. We see from Fig. 6.5 that the sample autocorrelations r_k decay rapidly, with only the first six outside the $\pm 2/\sqrt{n}$ limits, and $r_k \simeq 0$ beyond $k = 12$. This completes the first phase of the provisional model identification, namely $d = 1$.

Looking again at Fig. 6.5a we note that there is no sharp cut-off in the correlogram, suggesting that an MA model for the differenced data would be inappropriate. However, the partial correlogram, Fig. 6.5b, does show a sharp cut-off after lag one, suggesting an AR(1) model. We therefore identify (correctly) an ARIMA (1,1,0) model for the original data.

The above example does not provide a severe test of the model identification procedure for several reasons: the series is relatively long, leading to quite precise estimates of the correlogram and partial correlogram; there is a true underlying ARIMA process; and knowing the answer before we start is a help!

With shorter series, and correspondingly wider limits $\pm 2/\sqrt{n}$, or with true autoregressive or moving average parameter values close to zero, any cut-off in the correlogram or partial correlogram is more easily obscured by sampling fluctuations. With real data for which an ARIMA process is at best an approximate model, there will be a further element of compromise in choosing between alternative approximations within the ARIMA class. With regard to the first point, Fig. 6.6 shows the correlogram and partial correlogram for a second simulation, in which

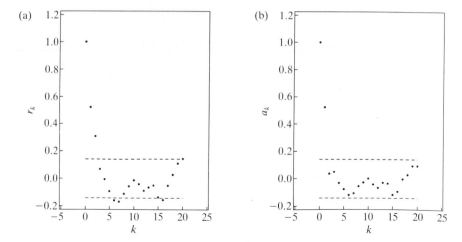

Fig. 6.6. The correlogram and partial correlogram of a simulated realization of an AR(1) process, $Y_t = 0.5Y_{t-1} + Z_t$, $n = 200$ observations. (a) Correlogram. (b) Partial correlogram.

the series is again of length 200 but the underlying process is now an AR(1) with autoregressive parameter 0.5, rather than 0.7. The exponential decay in the correlogram is correspondingly more rapid, and only the first two sample autocorrelations lie outside the $\pm 2/\sqrt{n}$ limits. Interpreting this as a cut-off would lead to the incorrect identification of an MA(2) model. Certainly, the cut-off in the partial correlogram is the more clear-cut, and a cautious strategy might be to favour an AR(1) model, with an MA(2) as a second contender. Alternatively, we might adopt an ARMA (1, 2) as the provisional model, but use the more formal methodology of Sections 6.3 and 6.4 to decide whether either the autoregressive or the moving average components should be dropped from the model.

6.3 Estimation

When ARMA models were first developed, limitations on available computing power necessitated a range of more or less *ad hoc* approaches to parameter estimation for particular models. Now, the method of maximum likelihood provides a unified, and entirely practicable, approach to parameter estimation for ARMA(p, q) processes of arbitrary order.

   In discussing parameter estimation we shall incorporate a non-zero mean, $\mu = E[Y_t]$, into our model, and assume that an observed series

$\{y_t : t = 1, \ldots, n\}$ is a realization of the ARMA (p, q) – process $\{Y_t\}$ defined by

$$Y_t - \mu = \sum_{i=1}^{p} \alpha_i (Y_{t-i} - \mu) + Z_t + \sum_{j=1}^{q} \beta_j Z_{t-j}.$$

This process involves $p + q + 2$ parameters: the α_i, the β_j, $\sigma^2 = \text{var}(Z_t)$ and $\mu = E[Y_t]$. Write $\alpha = (\alpha_1, \ldots, \alpha_p)'$, $\beta = (\beta_1, \ldots, \beta_q)'$, $\mathbf{y} = (y_1, \ldots, y_n)'$, $\mathbf{Y} = (Y_1, \ldots, Y_n)'$ and define a matrix $V(\alpha, \beta)$ such that

$$\text{var}(\mathbf{Y}) = \sigma^2 V(\alpha, \beta).$$

The elements of $V(\alpha, \beta)$ are proportional to the autocorrelation coefficients of $\{Y_t\}$ – proportional to rather than equal to, because we have chosen the parameter σ^2 to represent the variance of $\{Z_t\}$, not the variance of $\{Y_t\}$.

We make the additional assumption that the Z_t are Normally distributed, from which it follows that $\mathbf{Y}$ has a multivariate Normal distribution and the log-likelihood for μ, σ^2, $\alpha = (\alpha_1, \ldots, \alpha_p)'$ and $\beta = (\beta_1, \ldots, \beta_q)'$ is

$$L\{\mu, \sigma^2, \alpha, \beta\} = -\tfrac{1}{2}[n \log \sigma^2 + \log V(\alpha, \beta)$$
$$+ (\mathbf{y} - \mu\mathbf{1})'\{V(\alpha, \beta)\}^{-1}(\mathbf{y} - \mu\mathbf{1})/\sigma^2], \quad (6.3.1)$$

where $\mathbf{1}$ denotes a vector with all its elements equal to unity. We now proceed exactly as in Section 5.5. For given values of α and β, the maximum likelihood estimates of μ and σ^2 are

$$\hat{\mu}(\alpha, \beta) = [\mathbf{1}'\{V(\alpha, \beta)\}^{-1}\mathbf{y}]/[\mathbf{1}'\{V(\alpha, \beta)^{-1}\mathbf{1}] \quad (6.3.2)$$

and

$$\sigma^2(\alpha, \beta) = n^{-1}\{\mathbf{y} - \hat{\mu}(\alpha, \beta)\mathbf{1}\}'/\{V(\alpha, \beta)\}^{-1}\{\mathbf{y} - \hat{\mu}(\alpha, \beta)\mathbf{1}\}. \quad (6.3.3)$$

Back-substitution of eqns (6.3.2) and (6.3.3) into eqn. (6.3.1) gives the reduced log-likelihood

$$L_0(\alpha, \beta) = -\tfrac{1}{2}[n \log \sigma^2(\alpha, \beta) + \log V(\alpha, \beta)] \quad (6.3.4)$$

from which, in principle, maximum likelihood estimates $\hat{\alpha}$ and $\hat{\beta}$ can be obtained by numerical maximization.

An important qualification to the above is that each evaluation of the reduced log-likelihood involves inverting an n by n matrix. In the context of Chapter 5, we could assume that in most practical applications, n would be small enough for this not to be a major problem. However, we now envisage fitting a model to a single, relatively long series and the matrix inversion may prove to be prohibitive. Fortunately, more efficient algorithms for computing the log-likelihood function are widely available. See, for example, Newbold (1974) or Jones (1980). In order to demonstrate the close relationship between maximum likelihood and the more intuitive method of least squares, we now develop an explicit solution in one special case.

We consider the AR(1) process,

$$(Y_t - \mu) = \alpha(Y_{t-1} - \mu) + Z_t, \qquad (6.3.5)$$

and suppose that we wish to estimate μ, σ^2 and α from an observed series $\{y_t : t = 1, \ldots, n\}$. A simple-minded approach might be the following:

(a) μ represents the mean value of each Y_t, so we estimate it by $\hat{\mu} = \bar{y}$, the sample mean of the data,
(b) α represents the first autocorrelation coefficient of $\{Y_t\}$, so we estimate it by $\hat{\alpha} = r_1$, the first sample autocorrelation coefficient,
(c) given $\hat{\mu}$ and $\hat{\alpha}$, we can construct residuals

$$z_t = (y_t - \hat{\mu}) - \hat{\alpha}(y_{t-1} - \hat{\mu}) \quad : \quad t = 2, \ldots, n,$$

and estimate $\sigma^2 = \mathrm{var}(Z_t)$ by the residual mean square

$$\left(\sum_{t=2}^{n} z_t^2 \right) \Big/ (n - 1).$$

A slightly more formal approach, by analogy with classical linear regression, would be to estimate μ and α to minimize the residual sum of squares,

$$S(\mu, \alpha) = \sum_{t=2}^{n} \{(y_t - \mu) - \alpha(y_{t-1} - \mu)\}^2. \qquad (6.3.6)$$

Note that in eqn. (6.3.6) the summation begins with $t = 2$ rather than with $t = 1$ because y_0 is not observed. To obtain the form of the resulting 'least squares', estimates, we differentiate $S(\mu, \alpha)$. This gives

$$\frac{\partial S}{\partial \mu}(\mu, \alpha) = 2 \sum_{t=2}^{n} \{(y_t - \mu) - \alpha(y_{t-1} - \mu)\}(\alpha - 1)$$

$$= 2(\alpha - 1)\left\{ \sum_{t=2}^{n} (y_t - \alpha y_{t-1}) + (\alpha - 1)(n - 1)\mu \right\}, \qquad (6.3.7)$$

and

$$\frac{\partial S}{\partial \alpha}(\mu, \alpha) = -2 \sum_{t=2}^{n} \{(y_t - \mu) - \alpha(y_{t-1} - \mu)\}(y_{t-1} - \mu)\}.$$

$$= -2\left\{ \sum_{t=2}^{n} (y_t - \mu)(y_{t-1} - \mu) - \alpha \sum_{t=2}^{n} (y_{t-1} - \mu)^2 \right\}. \qquad (6.3.8)$$

Setting both partial derivatives equal to zero gives the least squares estimates $\hat{\mu}$ and $\hat{\alpha}$. From eqn. (6.3.7),

$$\hat{\mu} = \sum_{t=2}^{n} (y_t - \hat{\alpha} y_{t-1})/\{(1 - \hat{\alpha})(n - 1)\}$$

$$= \left\{ -\hat{\alpha} y_1 + (1 - \hat{\alpha}) \sum_{t=2}^{n-1} y_t + y_n \right\} \Big/ \{(1 - \hat{\alpha})(n - 1)\}.$$

From eqn. (6.3.8),

$$\hat{\alpha} = \sum_{t=2}^{n} (y_t - \hat{\mu})(y_{t-1} - \hat{\mu}) \Big/ \sum_{t=2}^{n} (y_{t-1} - \hat{\mu})^2.$$

Both $\hat{\mu}$ and $\hat{\alpha}$ are approximately equal to the simple-minded estimates $\hat{\mu} = \bar{y}$ and $\hat{\alpha} = r_1$, the difference arising only through the slightly different treatment of end-values y_1 and y_n. Analogy with classical regression would suggest estimating σ^2 by the residual mean square with degrees of freedom adjusted for estimation of μ and α,

$$\hat{\sigma}^2 = \sum_{t=2}^{n} \{(y_t - \hat{\mu}) - \hat{\alpha}(y_{t-1} - \hat{\mu})\}^2/(n - 3),$$

again approximately equal to the simple-minded estimate previously suggested.

Now, consider maximum likelihood estimation. We first derive the form of the log-likelihood function conditional on the first observation, y_1. From eqn. (6.3.5), it is clear that the conditional distribution of each Y_t, given the values y_s of all preceding Y_s, depends only on y_{t-1}. It follows that the joint probability density of $Y_2, \ldots, Y_n$ conditional on y_1 can be expressed as

$$f(y_2, \ldots, y_n \mid y_1) = \prod_{t=2}^{n} g(y_t \mid y_{t-1}),$$

each $g(\cdot)$ representing the conditional density of Y_t given y_{t-1}. Furthermore, if the Z_t are Normally distributed with mean zero and variance σ^2, then the conditional distribution of Y_t given y_{t-1} is Normal with mean $\mu + \alpha(y_{t-1} - \mu)$ and variance σ^2. Hence,

$$g(y_t \mid y_{t-1}) = (2\pi\sigma^2)^{-1/2} \exp[-\{y_t - \mu - \alpha(y_{t-1} - \mu)\}^2/(2\sigma^2)]$$

and

$$f(y_2, \ldots, y_n \mid y_1)$$

$$= (2\pi\sigma^2)^{-(n-1)/2} \exp\left[-\sum_{t=2}^{n} \{y_t - \mu - \alpha(y_{t-1} - \mu)\}^2/(2\sigma^2) \right]$$

Taking logarithms in the above expression gives the conditional log-likelihood for σ^2, μ and α as

$$L_c(\sigma^2, \mu, \alpha) = -\tfrac{1}{2}(n-1)\ln\sigma^2 - \sum_{t=2}^{n}\{y_t - \mu - \alpha(y_{t-1} - \mu)\}^2/(2\sigma^2) \quad (6.3.9)$$

Clearly, to maximize eqn. (6.3.9) with respect to μ and α, we need to minimize the quantity

$$\sum_{t=2}^{n}\{(y_t - \mu) - \alpha(y_{t-1} - \mu)\}^2,$$

which is precisely $S(\mu, \alpha)$ as defined in eqn. (6.3.6). Thus, conditional maximum likelihood and least squares estimation coincide. Also, differentiating (6.3.9) with respect to σ^2 gives

$$\frac{\partial L_c}{\partial\sigma^2}(\sigma^2, \mu, \alpha) = \tfrac{1}{2}(n-1)/\sigma^2 + S(\mu, \alpha)/2\sigma^4). \quad (6.3.10)$$

Setting eqn. (6.3.10) equal to zero then gives the conditional maximum likelihood estimator for σ^2 as

$$S(\hat{\mu}, \hat{\alpha})/(n-1) = \sum_{t=2}^{n} z_t^2/(n-1),$$

which differs from the least squares estimate σ^2 only with regard to the appearance of $n-1$ rather than $n-3$ in the denominator.

Finally, we consider full maximum likelihood estimation for the AR(1) process. To obtain the log-likelihood function, we need to add to the conditional log-likelihood (6.3.9) the logarithm of the marginal probability density of Y_1. By assumption, Y_1 is Normally distributed, with mean μ. By stationarity, the Y_t have a common variance, v^2 say. Then, from eqn. (6.3.5) we deduce that

$$v^2 = \alpha v^2 + \sigma^2,$$

which gives $v^2 = \sigma^2/(1 - \alpha^2)$. The marginal probability density of Y_1 is therefore

$$f_1(y_1) = \{2\pi\sigma^2/(1-\alpha^2)\}^{-1/2}\exp-\{(1-\alpha^2)(y_1-\mu)^2/(2\sigma^2)\},$$

and the log-likelihood function is

$$L(\sigma^2, \mu, \alpha) = L_c(\sigma^2, \mu, \alpha) - \tfrac{1}{2}\log\sigma^2 + \tfrac{1}{2}\log(1-\alpha^2)$$
$$- (1-\alpha^2)(y_1-\mu)^2/(2\sigma^2). \quad (6.3.11)$$

The maximum likelihood estimates cannot be written down explicitly. However, comparing eqns (6.3.11) and (6.3.9), we see that when n is large, the difference $L(\sigma^2, \mu, \alpha) - L_c(\sigma^2, \mu, \alpha)$ will be negligible by comparison with $L_c(\sigma^2, \mu, \alpha)$. Thus, maximum likelihood and conditional maximum likelihood or least squares estimates will usually differ only slightly. An exception is if α is very close to 1 or -1, in which case the term $\frac{1}{2}\log(1 - \alpha^2)$ in $L(\sigma^2, \mu, \alpha)$ can make a non-negligible contribution.

To summarize: maximum likelihood estimation provides a unified approach to parameter estimation for ARMA-processes. It has been implemented in published algorithms and is widely available in packages. In most cases of practical interest, maximum likelihood estimation is a minor variation on the more intuitively based method of least squares estimation. However, when the two methods differ non-trivially, maximum likelihood estimation is the more efficient, and is therefore to be preferred.

Example 6.2. Parameter estimation for simulated data
One widely available package which implements maximum likelihood and least squares estimation for ARMA processes is GENSTAT (Payne 1987). We used this package to fit an AR(1) model to the first differences of the simulated data from Fig. 6.1. Maximum likelihood and least squares estimates of the model parameters, with estimated standard errors, are given in Table 6.1. Note that the maximum likelihood and least squares estimates are approximately equal, as is to be expected for such a long series. Also, the standard errors indicate that both sets of estimates are in good statistical agreement with the true parameter values.

6.4 Diagnostic checking

After fitting a provisional ARMA model, we can assess its adequacy in various ways. The usual approach is to extract from the data a sequence of residuals to correspond to the underlying, but unobservable, white noise sequence, and to check that the statistical properties of these residuals are indeed consistent with white noise. Box and Jenkins (1970)

Table 6.1. Parameter estimation for an AR(1) model fitted to the first differences of the simulated data from Example 6.1

		Maximum likelihood		Least squares	
Parameter	True value	Estimate	Standard error	Estimate	Standard error
μ	0.0	0.154	0.231	0.153	0.234
α	0.7	0.698	0.051	0.702	0.051
σ	1.0	0.987		0.987	

advocate combining analysis of residuals with a procedure which they call 'trial overfitting'. In this procedure, the provisional model is compared with models containing an additional autoregressive or moving average parameter, and an assessment is made of whether this results in a significantly improved fit. A third approach, which appears to be little used, although the author believes it has considerable merit, is to compare the fitted ARMA spectrum with a non-parametric spectral estimate.

6.4.1 Residuals

All ARMA processes include in their definition an embedded white noise sequence $\{Z_t\}$. The *residuals* $\{z_t : t = 1, \ldots, n\}$ are obtained by substituting into the defining equation (6.3.1) the estimated values of all autoregressive and moving average parameters and the observed time series $\{y_t : t = 1, \ldots, n\}$, and solving the resulting set of equations for $\{z_t : t = 1, \ldots, n\}$.

Extraction of the residual sequence is particularly transparent in the case of the AR(p) process. In this case

$$Y_t - \mu = \sum_{j=1}^{p} \alpha_j (Y_{t-j} - \mu) + Z_t,$$

and the residuals are

$$z_t = (y_t - \hat{\mu}) - \sum_{j=1}^{p} \hat{\alpha}_j (y_{t-j} - \hat{\mu}) \quad : \quad t = p + 1, \ldots, n.$$

Note that z_t is undefined for $t \leq p$.

In the case of an MA(q) process, the residuals are extracted recursively. Note that the defining equation of the process is

$$Y_t = \mu + Z_t + \sum_{j=1}^{q} \beta_j Z_{t-j},$$

or

$$Z_t = (Y_t - \mu) - \sum_{j=1}^{q} \beta_j Z_{t-j}.$$

Set $z_t = 0$ for $t \leq 0$, and compute

$$z_1 = (y_1 - \hat{\mu}),$$
$$z_2 = (y_2 - \hat{\mu}) + \beta_1 Z_1$$

and so on until, for $t > q$,

$$z_t = (y_t - \hat{\mu}) + \sum_{j=1}^{q} \beta_j z_{t-j}.$$

A cautious strategy would be to discard the z_t for $t \leq q$.

Finally, for an ARMA (p, q) process, the defining equation gives

$$Z_t = (Y_t - \mu) - \sum_{i=1}^{p} \alpha_i (Y_{t-i} - \mu) - \sum_{j=1}^{q} \beta_j Z_{t-j}.$$

Setting $z_t = 0$ for $t \leqslant p$, we can then extract the later z_t as

$$z_t = (y_t - \hat{\mu}) - \sum_{i=1}^{p} \alpha_i (y_{t-1} - \hat{\mu}) - \sum_{j=1}^{q} \hat{\beta}_j z_{t-j}.$$

In this case, a cautious strategy is to discard the z_t for $t \leqslant \max(p, q)$.

6.4.2 Testing the residuals against white noise

The residual sequence $\{z_t\}$ can be subjected to any of the tests for white noise which we introduced in Chapter 2. These tests are not strictly valid because they take no account of the effects of parameter estimation. It is nevertheless sensible to compute both the correlogram and the cumulative periodogram of the residual series, and to assess these informally for compatibility with white noise.

6.4.3 Trial overfitting

Suppose that we have provisionally identified an ARMA(p, q) process as an appropriate model and that we have estimated its parameters by the method of maximum likelihood. Write L_0 for the maximized value of the log-likelihood. Now, repeat the fitting procedure for ARMA$(p + 1, q)$ and ARMA $(p, q + 1)$ models, obtaining maximized log-likelihood values L_1 and L_2, respectively. According to the standard theory of generalized likelihood ratio testing, if the provisional model is correct each of the statistics $2(L_1 - L_0)$ and $2(L_2 - L_0)$ is distributed as chi-squared on one degree of freedom.

Trial overfitting is attractive in its conceptual simplicity and in its apparent objectivity. However, it does tend to be self-servicing, in the sense that the likelihood ratio criterion, which is based on Normality assumptions, is heavily influenced by the correlation structure of the data, and these are precisely the aspects of the data which have been used in identifying the provisional model.

6.4.4 Spectral checking

In Chapter 4, we showed how to estimate a spectrum from an observed stationary time series without making parametric assumptions about the underlying random process. In particular, if $f(\omega)$ denotes the true spectrum and $\hat{f}(\omega)$ a spectral estimate based on a $(2p + 1)$-point simple moving average of periodogram ordinates, then

$$\hat{f}(\omega) \sim f(\omega)\chi^2_{2(2p+1)}/\{2(2p+1)\}. \tag{6.4.1}$$

It follows that if we define critical values c_1 and c_2 such that

$$P\{\chi^2_{2(2p+1)} \leqslant c_1\} = P\{\chi^2_{2(2p+1)} \geqslant c_2\} = \tfrac{1}{2}\alpha,$$

then for each Fourier frequency ω,

$$P[f(\omega)c_1/\{2(2p+1)\} \leqslant \hat{f}(\omega) \leqslant f(\omega)c_2/\{2(2p+1)\}] = 1 - \alpha. \tag{6.4.2}$$

In other words, the range of values from $f(\omega)c_1/\{2(2p+1)\}$ to $f(\omega)c_2/\{2(2p+1)\}$ defines a $100(1-\alpha)\%$ tolerance interval for the spectral estimate $\hat{f}(\omega)$. Of course, the true spectrum $f(\omega)$ is unknown, but we can compute tolerance limits using the fitted ARMA spectrum in place of $f(\omega)$, and thereby assess whether the spectral estimate $\hat{f}(\omega)$ is consistent with the fitted ARMA model.

As already noted in Chapter 2, although the correlogram and the periodogram both encapsulate the second-order properties of an observed time-series, they often highlight different aspects of these properties. To this extent at least, spectral checking has the attractive property that it complements, rather than duplicates, the statistical procedures used to identify and fit the provisional ARMA model.

Example 6.3. Diagnostic checking for simulated data
In this example, we again use the simulated data from Fig. 6.1 to illustrate the statistical methodology. We first extract from the data the residual series using the maximum likelihood estimates of the parameters in the fitted ARIMA(1, 1, 0) model, rather than the true parameter values. Note that this series is of length 198, since we lose one of the original 200 observations because of the differencing, and a second because of the autoregressive term.

Figure 6.7 shows the correlogram of the residual series. All of the sample autocorrelations $r_k : k = 1, \ldots, 20$ fall within the $\pm 2/\sqrt{n}$ limits, indicating compatibility with white noise. Figure 6.8 shows the cumulative periodogram with the 5% critical value for a test of white noise using D indicated by a pair of parallel lines. The cumulative periodogram lies entirely between these lines, again indicating compatibility with white noise. Both graphical displays therefore support the adequacy of the fitted model.

In this case, trial overfitting consists of comparing the maximized log-likelihood for the AR(1) model fitted to the first differenced data with the corresponding maximized log-likelihoods from AR(2) and ARMA(1, 1), 'overfits'. Table 6.2 lists the three values of the 'deviance'. defined to be minus twice the maximized log-likelihood. Both the AR(2)

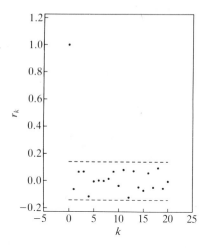

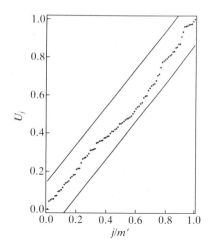

Fig. 6.7. The correlogram of the residual series obtained after fitting an ARIMA$(1, 1, 0)$ model to the data from Fig. 6.1. The dashed horizontal lines correspond to the limits $\pm 2/\sqrt{n}$.

Fig. 6.8. The cumulative periodogram of the residual series obtained after fitting an ARIMA$(1, 1, 0)$ model to the data from Fig. 6.1. The pair of parallel lines correspond to the 5% critical value of the statistic D.

and ARMA$(1, 1)$ models necessarily give a smaller deviance, i.e. a larger log-likelihood, than does the AR(1) model. However, if the AR(1) model is correct, then the reduction in deviance between the AR(1) model and either the AR(2) or the ARMA$(1, 1)$ model should be distributed as χ_1^2. The observed reductions are 1.66 for the AR(2) and 1.46 for the ARMA $(1, 1)$, both clearly compatible with χ_1^2.

For a spectral check on the fitted model, we compute a spectral estimate as a discrete spectral average of order 5. Figure 6.9 shows this estimate on a logarithmic scale, together with the spectrum of the fitted model and pointwise upper and lower 5% tolerance limits derived from eqns (6.4.1) and (6.4.2). If the fitted model is correct, approximately 90% of the estimated spectral ordinates should lie within the tolerance

Table 6.2. Deviances associated with fits for the first differences of the simulated data from Example 6.1

Model	Deviance	Degree of freedom
AR(1)	194.44	197
AR(2)	192.78	196
ARMA(1, 1)	192.98	196

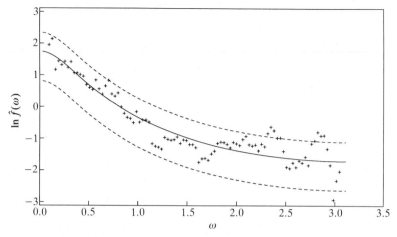

Fig. 6.9. A spectral check on the model fitted to the first difference of the data from Fig. 6.1. +, Non-parametric spectral estimate computed as a discrete spectral average of order 5. ———, Spectrum of the fitted AR(1) model; – – –, pointwise 90% tolerance limits for the non-parametric estimate, assuming the fitted spectrum.

limits; in fact, 83 out of 94 lie clearly within the limits, and a further four lie almost exactly on the upper limit. Thus, the spectral check again supports the adequacy of the fitted model.

6.5 Case studies

We have already acknowledged that the simulated example used here to illustrate the fitting of ARIMA models to data shows the methodology in the best possible light. We therefore conclude this chapter with two examples involving real data.

Example 6.4. Fitting an ARIMA model to the log-wool-price-ratio
Figure 6.10 shows the series $y_t = \log$ (19 μm price/floor price) derived from the data of Example 1.3. In this example, we ignore the gaps in the sequence of weekly markets and treat the data as equally spaced. The data are clearly non-stationary, and the first step in identifying a possible ARIMA model for them is to take first differences. The differenced series is plotted in Fig. 6.11. There appears to have been an increase in variability after about the first 100 observations. How we react to this depends on our purpose in analysing the data. If our objective is an understanding of the mechanisms which drive the data, the change in

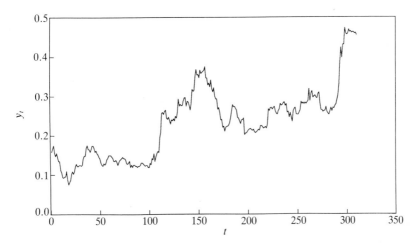

Fig. 6.10. The log-wool-price-ratio series, $y_t = \log$ (19 μm price/floor price).

behaviour is of inherent interest and we should seek to explain it, possibly in terms of external forces which might have led to the change in behaviour in early 1979. If, on the other hand, our aim is to develop a model for forecasting future values of y_t, it would be unwise to allow the model to be influenced by data from a time at which the behaviour of the data was qualitatively different from the present. In anticipation of the next chapter, we take the latter view, discard the first 100 observations and search for a model for the remaining 209.

Figure 6.12 shows the correlogram of the series of 209 first differences.

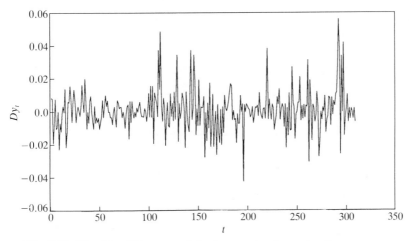

Fig. 6.11. The first difference of the log-wool-price-ratio, $Dy_t = y_t - y_{t-1}$.

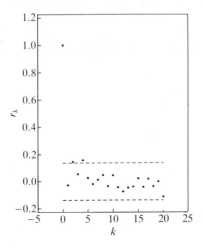

Fig. 6.12. The correlogram of $n = 209$ first differences Dy_t, after discarding the first 100 observations from Fig. 6.1. The dashed horizontal lines correspond to the limits $\pm 2/\sqrt{n}$.

Although the second and fourth sample autocorrelations are slightly greater than the $2/\sqrt{n}$ upper limit, there is no discernible pattern in the correlogram, and we conclude that the first differenced data are compatible with white noise. In fact, the sample mean and variance of the 209 first differences are $0.001\,556$ and $0.000\,190$ respectively, suggesting a model for the original data of the form

$$Y_t = 0.001\,556 + Y_{t-1} + Z_t, \tag{6.4.3}$$

where $\{Z_t\}$ is a zero-mean white noise sequence with variance $0.000\,190$. The standard error of the sample mean is approximately $0.000\,915$, and the t-statistic to test for a non-zero mean is 1.63 on 208 degrees of freedom, not quite significant at the 10% level. Whether or not we choose to retain a non-zero mean in eqn. (6.4.3) has implications for forecasting, which we shall pursue in the next chapter.

Example 6.5. Fitting an ARIMA model to an LH series
In Example 4.9 we fitted a parametric model to LH data from the late follicular phase of the menstrual cycle. This model assumed, amongst other things, that the LH data in the early follicular phase followed an AR(1) model. For comparison, we now fit an ARIMA model directly to the early follicular phase series of $n = 48$ observations which is the second of the three series constituting Example 1.2.

Figure 1.3b showed a time plot of this series, which appears to be stationary. Figure 6.13 shows the correlogram and partial correlogram,

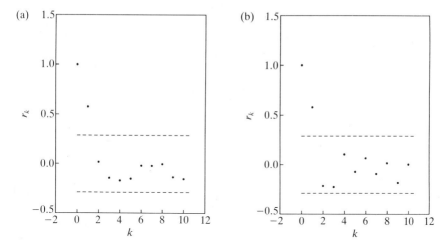

Fig. 6.13. The correlogram and partial correlogram of the early follicular phase LH time series. The dashed horizontal lines correspond to the limits $\pm 2/\sqrt{n}$. (a) Correlogram. (b) Partial correlogram.

with the usual $\pm 2/\sqrt{n}$ limits. These immediately highlight the difficulties of interpretation which arise with short series. Viewed strictly in relation to the $\pm 2/\sqrt{n}$ limits, both the correlogram and partial correlogram could be said to exhibit a cut-off after lag one. The cut-off appears sharper in the partial correlogram, suggesting an AR(1) model as previously assumed. However, on this evidence alone we cannot rule out an MA(1).

One way to attempt to resolve the uncertainty is to fit the two models, also an ARMA(1, 1) model which includes them both, and use the deviances from the three models as a basis for choosing amongst them. The deviances for the three models are 9.48 for the AR(1), 10.20 for the MA(1) and 9.30 for the ARMA(1, 1). Neither the reduction in deviance of 0.18 in moving from the AR(1) to the ARMA(1, 1) model nor the reduction of 0.90 in moving from the MA(1) to the ARMA(1, 1), is significant relative to χ_1^2, and there appears to be no justification for including both the autoregressive and the moving average parameters. The AR(1) model remains a reasonable choice since it has smaller deviance that the MA(1). However, in truth the series is too short to discriminate effectively between these two models.

Adopting the AR(1) model, we obtain maximum likelihood estimates $\hat{\mu} = 2.41$ with standard error 0.15 for the mean and $\hat{\alpha} = 0.57$ with standard error 0.12 for the autoregressive parameter, i.e. the fitted model is

$$Y_t - 2.41 = 0.57(Y_{t-1} - 2.41) + Z_t,$$

where $\{Z_t\}$ is zero-mean white noise with estimated variance 0.2061.

Diagnostic checks on the AR(1) model are summarized in Fig. 6.14. These consist of the correlogram of the residuals, the cumulative periodogram of the residuals, and a comparison between the fitted autoregressive spectrum and a discrete spectral average of order three. All three checks suggest an adequate fit.

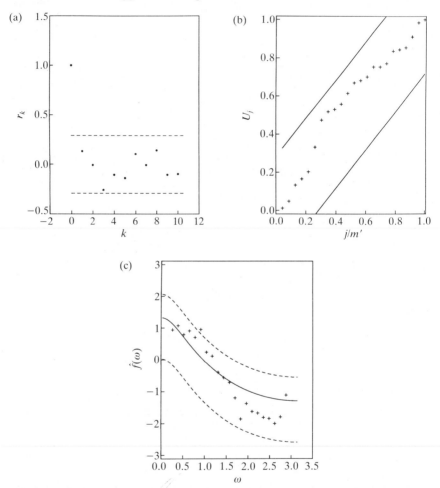

Fig. 6.14. Diagnostic checks on the AR(1) model fitted to the early follicular LH time series. (a) Correlogram of residuals, with dashed horizontal lines corresponding to the limits $\pm 2/\sqrt{n}$. (b) Cumulative periodogram of residuals, with parallel lines to correspond to the 5% critical value of the statistic D. (c) A spectral check, $+$, Non-parametric spectral estimate computed as a discrete spectral average of order 3. ———, Spectrum of the fitted AR(1) model; $---$, pointwise 90% tolerance limits for the non-parametric estimate, assuming the fitted spectrum.

An approximate 95% confidence interval for α, constructed as $\hat\alpha$ plus or minus two standard errors, extends from 0.33 to 0.81. By comparison, in Example 4.9 we estimated $\hat\alpha = 0.76$ in the course of fitting a more complex model to the two late follicular phase series from Example 1.2. That the two analyses, based on different data, are consistent with each other adds credibility to the model as a reasonable description of the data.

6.6 Further reading

Box and Jenkins (1970) give a very detailed account of methods for fitting ARIMA processes to time-series data. Tunnicliffe-Wilson (1989) describes a variant of maximum likelihood estimation which gives a useful improvement over maximum likelihood when the constant mean value μ is replaced by a regression function, for example a polynomial in time (see also Section 5.7). Jones (1980) gives an implementation of maximum likelihood based on the state space representation of an ARMA process, which can accommodate missing values and superimposed measurement error, i.e. the data are generated by a model of the form $D_t = Y_t + E_t$, where $\{Y_t\}$ is an ARMA process and $\{E_t\}$ is white noise. Note that in this model, $\{E_t\}$ is distinct from the white noise process $\{Z_t\}$ embedded in $\{Y_t\}$.

7
Forecasting

7.1 Preamble

In this chapter, we discuss how an observed time-series $y_1, \ldots, y_n$ can be used to make predictions of its future behaviour. As an example, consider the data of Example 1.6 on the growth of colonies of *paramecium aurelium*. Two questions which might be asked of these data are:

(a) what is the *average* size of a colony as a function of time since the start of the experiment?
(b) what is the best prediction of the future size of a *particular* colony, given its growth to date?

The analysis of these data described in Chapter 5 was directed at the first of these questions, whereas this chapter is concerned with the second. Specifically, we consider the problem of predicting, or forecasting, a future value, y_{n+k} say, from an observed series $y_1, \ldots, y_n$.

Before discussing specific forecasting methods, we make the point that forecasting involves extrapolation in the presence of random variation, and as such is bound to be difficult in practice. The difficulties are threefold. Firstly, we need to be confident that the mechanisms generating the historical data will continue to apply in the future. Secondly, even if we can assume a stable generating mechanism, we need to express this mechanism in the form of a statistical model for the data. The difficulty here is that different models may fit the historical data equally well, yet give quite different predictions of the future. Thirdly, even supposing we have identified the correct statistical model, the statistical uncertainty inherent in any estimates of the model parameters will be magnified as predictions are made progressively further into the future.

Perhaps the simplest illustration of this last point is the well-known problem of prediction for a simple linear regression model. We assume that the data $y_1, \ldots, y_n$ are generated by the model

$$y_t = \alpha + \beta(t - \bar{t}) + Z_t \quad : \quad t = 1, \ldots, n \qquad (7.1.1)$$

where $\bar{t} = n^{-1} \sum_{t=1}^{n} t = (n+1)/2$ and Z_t is a Normally distributed white noise sequence with variance σ^2. The least squares or maximum likelihood estimates of α and β are given by

$$\hat{\alpha} = \bar{y}; \qquad \hat{\beta} = \sum_{t=1}^{n} y_t (t - \bar{t}) \bigg/ \sum_{t=1}^{n} (t - \bar{t})^2,$$

where $\bar{y} = n^{-1} \sum_{t=1}^{n} y_t$. A sensible point forecast of a future observation is then

$$\hat{\alpha} + \hat{\beta}\{(n + k) - \bar{t}\}. \qquad (7.1.2)$$

For the model (7.1.1), $\hat{\alpha}$ and $\hat{\beta}$ are uncorrelated, with variances

$$\text{var}(\hat{\alpha}) = \sigma^2/n; \qquad \text{var}(\hat{\beta}) = \sigma^2 \bigg/ \left\{ \sum_{t=1}^{n} (t - \bar{t})^2 \right\}^2.$$

It follows that the variance of the expression (7.1.2) is

$$V(k) = \text{var}(\hat{\alpha}) + (n + k - \bar{t})^2 \, \text{var}(\hat{\beta})$$

$$= \sigma^2 \left[n^{-1} + (n + k - \bar{t})^2 \bigg/ \left\{ \sum_{t=1}^{n} (t - \bar{t})^2 \right\}^2 \right]. \qquad (7.1.3)$$

Now, in eqn. (7.1.3), $n + k - \bar{t} = k + (n - 1)/2$, from which it is clear that $V(k)$ increases with k.

We can pursue the linear regression example a little further, to make a simple but important point. The quantity $V(k)$ in eqn. (7.1.3) is the variance of the value of the fitted regression line at time $t = n + k$. This quantity can be used to assess the precision with which we can estimate the *expected* value of the random variable Y_{n+k}. If we want to predict the *observed* value y_{n+k} of Y_{n+k}, the best point prediction is again

$$\hat{y}_{n+k} = \hat{\alpha} + \hat{\beta}\{(n + k) - \bar{t}\}$$

but its variance is now

$$\text{var}(\hat{y}_{n+k}) = \sigma^2 + V(k),$$

the additional σ^2 representing the variability of Y_{n+k} about its expected value.

In a biological context, this distinction between extrapolation of a trend and prediction of a future observation arises frequently in growth studies like the one described in Example 1.6, for which both questions might be relevant. In other studies this is not so. For example, in the case of the wool price data of Example 1.3, only the prediction of actual prices would seem to be relevant, because there is no sense in which the mechanism which determines the sequence of measurements can be replicated. In the remainder of this chapter, we shall use these data as a vehicle for illustrating and comparing a number of different forecasting methods. Specifically, we shall illustrate the performance of various methods for predicting values of the log-wool-price-ratio series,

$$Y_t = \log \text{(price paid/floor price)}$$

for fine grade (19 μm nominal thickness) wool.

Throughout the chapter, we shall use $Y_n(k)$ to denote the prediction for Y_{n+k} from observed values y_t up to an including $t = n$, and restrict our attention to *short-term* forecasting, i.e. small values of k. In Section 7.2 we consider the use of polynomial trends for prediction and conclude that this is unsatisfactory, in part because the fitted trends are unduly influenced by observations from the remote past. This leads on to a discussion of *exponential smoothing* in Section 7.3, which gives one way of automatically discounting the influence of the remote past. In Section 7.4 we give an outline of the Box and Jenkins approach to forecasting. This involves fitting ARMA models to historical data, possibly after a differencing transformation, and using the fitted model to suggest an optimal prediction formula. A much more detailed account is given in Box and Jenkins (1970).

7.2 Forecasting by extrapolation of polynomial trends

Given an observed time series $y_1, \ldots, y_n$ it is a straightforward task to fit a polynomial trend to the data by least squares, as described in Section 2.2, and to compute forecasts by extrapolation of the fitted polynomial. If the data are indeed generated by a polynomial trend with white noise deviations about the trend, this is a very reasonable approach to take. However, in practice the resulting forecasts will often be poor, for at least two reasons. Firstly, the assumed statistical model may be wrong. Secondly, if a high-degree polynomial is required in order to give a reasonable fit to the observed data, the extrapolation may be quite bizarre, tending rapidly to plus or minus infinity outside the time interval of the observed data.

Example 7.1. Extrapolation of polynomial trends for the log-wool-price-ratio.

In this example we construct forecasts for the log-wool-price-ratio as follows. Fit a polynomial trend of degree d to the first m data points by least squares, extrapolate the fitted trend to time $m + k$, and use this extrapolated value as the k-step ahead forecast $Y_m(k)$. To evaluate the performance of this method of forecasting, we repeat the exercise for each of $m = 200, 201, 202, \ldots, 300$ and compute the *root mean square error*,

$$\text{rmse} = \sqrt{\left\{ \sum_{m=200}^{300} (Y_m(k) - y_{m+k})^2 / 101 \right\}}. \tag{7.2.1}$$

Table 7.1. Root mean square errors for polynomial trend extrapolation
applied to the log-wool-price-ratio

		d							
		1	2	3	4	5	6	7	8
k	1	0.056	0.055	0.072	0.057	0.057	0.066	0.066	0.031
	2	0.058	0.058	0.078	0.063	0.065	0.079	0.081	0.041
	3	0.060	0.061	0.084	0.068	0.073	0.092	0.098	0.054
	4	0.061	0.064	0.089	0.074	0.081	0.107	0.116	0.068
	5	0.063	0.067	0.095	0.080	0.090	0.123	0.136	0.084
	6	0.064	0.069	0.100	0.086	0.099	0.140	0.157	0.102
	7	0.066	0.072	0.106	0.092	0.108	0.158	0.181	0.123
	8	0.067	0.074	0.111	0.098	0.118	0.177	0.207	0.146
	9	0.068	0.076	0.117	0.105	0.128	0.198	0.234	0.172
	10	0.068	0.078	0.122	0.111	0.139	0.220	0.265	0.200

Table 7.1 gives rmse values for each of $d = 1, 2, \ldots, 8$ and each of
$k = 1, 2, \ldots, 10$. The most obvious, and unsurprising feature of the table
is that for every value of d, $\mathrm{rmse}(d, k)$ increases with k, i.e. forecasts
become less accurate as their lead time increases. A second feature is that
for small values of k, $\mathrm{rmse}(d, k)$ changes relatively little with d, whereas
for large values of k, $\mathrm{rmse}(d, k)$ generally increases with d. This
emphasizes the point, previously made in Example 5.7, that long-range
extrapolation of high-degree polynomial fits is usually a bad idea.

It is only fair to point out that, with regard to the wool-price data, the
polynomial trend model is something of a 'straw man'. In other contexts,
polynomial trends may give an adequate description of historical data and
reasonable short-term forecasts. However, the erratic fluctuations of the
wool-price data are not atypical of the type of economic time series for
which forecasts are most often required.

One way to improve the performance of polynomial trend extrapolation for the wool-price data is to fit the trend using only relatively recent
data, rather than the whole observed series. The rationale for this is that
the underlying trend in the data may be well approximated locally by a
polynomial, even when the global fit is poor.

Example 7.2. Extrapolation of local linear trends for the log-wool-price-
ratio
To illustrate the use of locally fitted linear trends for forecasting the
log-wool-price-ratio, we proceed as follows. At time m, fit a straight line

Table 7.2. Root mean square errors for local linear trend extrapolation applied to the log-wool-price-ratio

		\(r\)			
		3	5	10	25
\(k\)	1	0.016	0.016	0.018	0.031
	2	0.025	0.022	0.024	0.037
	3	0.034	0.029	0.030	0.042
	4	0.042	0.036	0.036	0.048
	5	0.052	0.044	0.043	0.054
	6	0.062	0.052	0.050	0.059
	7	0.071	0.060	0.057	0.064
	8	0.081	0.067	0.064	0.070
	9	0.091	0.075	0.071	0.075
	10	0.100	0.084	0.079	0.079

to the r most recent observations, $y_{m-r+1}, y_{m-r+2}, \ldots, y_m$, extrapolate the fitted line to time $m + k$ and use this extrapolated value as the k-step ahead forecast $Y_m(k)$. As in Example 7.1, we evaluate the performance of the forecasts by rmse as defined at eqn. (7.2.1).

Table 7.2 gives rmse values for each of $r = 3$, 5, 10, 25, and $k = 1$, 2, ..., 10. As in Example 7.1, the rmse values increase with k, whatever the value of r used. For any particular value of k, the rmse values do not alter dramatically with r. A good choice of r needs to balance two considerations: small values of r will respond most effectively to genuine changes in the local trend, whereas large values will be most resistant to spurious apparent changes in trend which are in fact no more than random fluctuations. Finally, note that all four values of r give better results than global linear trend extrapolation for $k \leqslant 6$, but thereafter the global method does somewhat better. A reasonable explanation for this is that locally fitted linear trends are better suited to short-range extrapolation because the local trend varies over time in an irregular manner, whereas this same adaptability to *local* behaviour has a deleterious effect on long-range extrapolation.

7.3 Exponential smoothing

One reason for the failure of polynomial trend extrapolation is that observations from the remote past have a non-negligible influence on forecasts of the future. Local polynomial fitting is one way to overcome this. Another is to construct forecasts from weighted averages of the observations, with the most recent observations being given the largest weights. Thus, we consider forecasts of the form

$$Y_n(k) = \sum_{j=0}^{n-1} w_j y_{n-j} \qquad (7.3.1)$$

with $\sum_{j=0}^{n-1} w_j = 1$ and $w_0 \geqslant \cdots \geqslant w_{n-1}$. Note that this discussion exactly parallels the discussion in Section 2.2 concerning the relative merits of polynomial regression and weighted moving averages for smoothing historical data.

One possible choice of weights in eqn. (7.3.1) is to take

$$w_j = \alpha(1 - \alpha)^j \qquad (7.3.2)$$

for some $0 < \alpha < 1$. Strictly, the sum of these weights is not precisely 1, but $1 - (1 - \alpha)^n \approx 1$ for large n. Substitution of eqn. (7.3.2) into eqn. (7.3.1) gives

$$Y_n(k) = \alpha \sum_{j=0}^{n-1} (1 - \alpha)^j y_{n-j}$$

$$= \alpha y_n + (1 - \alpha) \left\{ \alpha \sum_{j=1}^{n-1} (1 - \alpha)^{j-1} y_{n-j} \right\}.$$

The term within braces in the above expression can be written as

$$\alpha \sum_{i=0}^{n-2} (1 - \alpha)^i y_{n-1-i} = Y_{n-1}(k),$$

so that

$$Y_n(k) = \alpha y_n + (1 - \alpha) Y_{n-1}(k). \qquad (7.3.3)$$

Equation (7.3.3) provides a simple formula for updating forecasts as each new observation becomes available.

This method of forecasting, due to C. C. Holt, is called *exponential smoothing* because the weights w_j defined by eqn. (7.3.2) decay exponentially. The value of α must be chosen by the user to reflect the desired influence of past observations on the forecast. Note in particular that if $\alpha = 1$, the forecast is simply the current observation, y_t.

Ostensibly, eqn. (7.3.3) provides the same forecasts $Y_n(k)$ for all values of k. In some circumstances, this may be quite sensible. However, in eqn (7.3.3) there is a concealed dependence of α on k, in the sense that for a given set of data, the optimal choice of α will depend on k.

Example 7.3. Exponential smoothing for the log-wool-price-ratio
As in Examples 7.1 and 7.2, we can use rmse values as defined at eqn.

Table 7.3. Root mean square errors for exponential smoothing applied to the wool price ratio

		α									
		0.1	0.2	0.3	0.4	0.5	0.6	0.7	0.8	0.9	1.0
k	1	0.036	0.027	0.022	0.018	0.016	0.015	0.014	0.014	0.013	0.013
	5	0.053	0.047	0.043	0.041	0.039	0.038	0.037	0.036	0.036	0.036
	10	0.069	0.066	0.063	0.062	0.060	0.060	0.059	0.058	0.058	0.058

(7.2.1) to assess the quality of forecasts obtained from exponential smoothing. Table 7.3 gives the results for each of $\alpha = 0.1, 0.2, \ldots, 1.0$, and $k = 1$, 5 and 10. For all three values of k, the rmse is smallest at or near $\alpha = 1.0$. The rmse is most responsive to changes in α for $k = 1$, decreasing by a factor of almost three as α increases from 0.1 to 1.0, whereas for $k = 5$ and $k = 10$, the rmse varies only slightly over the range of α. Finally, note that the exponential smoothing results with $\alpha = 1.0$ are better than any of the global or local trend extrapolation methods used in Examples 7.1 and 7.2.

It would be imprudent to draw general conclusions about the relative merits of different forecasting schemes from their performance on a single data set. Instead, a natural question to ask of a method such as exponential smoothing is whether there is some underlying statistical model for which the method is optimal, for example in the sense of minimizing the mean squared forecasting error. Put the other way round, if we can find a statistical model which adequately describes an observed time series, we could then derive a forecasting method which is optimal for the model in question. This is the essence of a very flexible approach to forecasting due to Box and Jenkins (1970).

7.4 The Box–Jenkins approach to forecasting

Box and Jenkins (1970) build a general forecasting methodology from the assumption that the time series in question, possibly after transformation and differencing, is generated by a stationary autoregressive moving average (ARMA) process. In this section, we give a brief outline of this methodology and refer the reader to the original work by Box and Jenkins (1970), or to Anderson (1976), for more detailed accounts.

Recall from Section 3.4 that an ARMA-process $\{Y_t\}$ is defined by the set of equations

$$Y_t = \sum_{j=1}^{p} \alpha_j Y_{t-j} + Z_t + \sum_{i=1}^{p} \beta_i Z_{t-i}, \qquad (7.4.1)$$

where $\{Z_t\}$ is white noise. Using the operator notation, we write this as

$$\phi(B)Y_t = \theta(B)Z_t, \qquad (7.4.2)$$

where $\phi(\cdot)$ and $\theta(\cdot)$ are polynomials, of degree p and q respectively, with $\phi(0) = \theta(0) = 1$ and with all roots of the polynomial equations $\phi(z) = 0$ and $\theta(z) = 0$ outside the unit circle. Recall also that $\{Y_t\}$ has a representation as a general linear process,

$$Y_t = \sum_{i=0}^{\infty} \theta_i Z_{t-i}. \qquad (7.4.3)$$

Now, suppose that we wish to construct a forecast, $Y_t(k)$, of the form

$$Y_t(k) = \sum_{j=0}^{t-1} w_j Y_{t-j}. \qquad (7.4.4)$$

Combining eqns (7.4.3) and (7.4.4) we obtain

$$Y_t(k) = \sum_{j=0}^{t-1} w_j \sum_{i=0}^{\infty} \theta_i Z_{t-j-i}$$

$$= \sum_{j=0}^{\infty} W_j Z_{t-j}, \qquad (7.4.5)$$

say. In eqn. (7.4.5) we are not concerned with the precise relationship between the W_j, w_j, and θ_i, but only with the fact that $Y_t(k)$ as defined by eqn. (7.4.4) can also be written in terms of the underlying white noise process $\{Z_t\}$ as a linear combination of current and past values.

We shall now evaluate the mean squared forecasting error. From eqns (7.4.3) and (7.4.5), this is

$$M = E[\{Y_{t+k} - Y_t(k)\}^2]$$

$$= E\left[\left\{\sum_{i=0}^{\infty} \theta_i Z_{t+k-i} - \sum_{j=0}^{\infty} W_j Z_{t-j}\right\}^2\right]$$

$$= E\left[\left\{\sum_{i=0}^{k-1} \theta_i Z_{t+k-i} + \sum_{i=k}^{\infty} (\theta_i - W_{i-k})Z_{t+k-i}\right\}^2\right],$$

$$= \sigma^2\left\{\sum_{i=0}^{k-1} \theta_i^2 + \sum_{i=k}^{\infty} (\theta_i - W_{i-k})^2\right\}, \qquad (7.4.6)$$

using the facts that the Z_t are mutually independent, each with mean zero and variance σ^2. Clearly, eqn. (7.4.6) is minimized by taking

$$W_i = \theta_{i+k} \quad : \quad i = 0, 1, \dots,$$

and the resulting forecast is

$$Y_t(k) = \sum_{j=0}^{\infty} W_j Z_{t-j}$$

$$= \sum_{j=0}^{\infty} \theta_{j+k} Z_{t-j}$$

$$= \sum_{i=k}^{\infty} \theta_i Z_{t+k-i}. \tag{7.4.7}$$

Comparing eqns (7.4.7) and (7.4.3), the latter re-written as

$$Y_{t+k} = \sum_{i=0}^{\infty} \theta_i Z_{t+k-i}, \tag{7.4.8}$$

we see that the optimal forecast $Y_t(k)$ has the following very simple interpretation: $Y_t(k)$ is evaluated from the definition of Y_{t+k}, except that the *future* values $Z_{t+1}, \dots, Z_{t+k}$ are set equal to zero. The forecast errors, $Y_{t+k} - Y_t(k)$, have an equally simple interpretation in terms of the underlying white noise process. From eqns (7.4.7) and (7.4.8),

$$Y_{t+k} - Y_t(k) = \sum_{i=0}^{k-1} \theta_i Z_{t+k-i}. \tag{7.4.9}$$

In particular, it follows from the conditions on the polynomials $\phi(\cdot)$ and $\theta(\cdot)$ in eqn. (7.4.2) that $\theta_0 = 1$, and therefore that

$$Y_{t+1} - Y_t(1) = Z_{t+1},$$

i.e. the one-step ahead forecast errors are the elements of the underlying white noise sequence $\{Z_t\}$.

To apply these results in practice, it is simpler to work directly with the explicit formula (7.4.1), rather than to evaluate the θ_i in terms of the original ARMA parameters α_j and β_i.

Example 7.4. $Y_t = \alpha Y_{t-1} + Z_t$
For this AR(1) process,

$$Y_{t+1} = \alpha Y_t + Z_{t+1},$$

so the optimal one-step-ahead forecast is

$$Y_t(1) = \alpha Y_t.$$

The optimal k-step-ahead forecasts for $k > 1$ can be evaluated in two different, but equivalent, ways. The first is to use appropriate forecasts in place of future values of Y_t. Thus, for the AR(1) process,

$$Y_{t+2} = \alpha Y_{t+1} + Z_{t+2},$$

leading to

$$Y_t(2) = \alpha Y_t(1) = \alpha^2 Y_t.$$

The second is to back-substitute from the defining equation of the process so as to eliminate future values of Y_t. Thus,

$$Y_{t+2} = \alpha Y_{t+1} + Z_{t+2}$$
$$= \alpha\{\alpha Y_t + Z_{t+1}\} + Z_{t+2}$$
$$= \alpha^2 Y_t + \alpha Z_{t+1} + Z_{t+2},$$

giving

$$Y_t(2) = \alpha^2 Y_t$$

as before.

Example 7.5. Exponential smoothing revisited

Suppose that the first difference of $\{Y_t\}$ is an MA(1) process, i.e. Y_t is ARIMA $(0, 1, 1)$,

$$Y_t = Y_{t-1} + Z_t + \beta Z_{t-1}. \tag{7.4.10}$$

Then,

$$Y_{t+1} = Y_t + Z_{t+1} + \beta Z_t.$$

Setting $Z_{t+1} = 0$ and using the fact that $Z_t = Y_t - Y_{t-1}(1)$, this gives the one-step-ahead forecast as

$$Y_t(1) = Y_t + \beta\{Y_t - Y_{t-1}(1)\}$$
$$= (1 + \beta)Y_t - \beta Y_{t-1}(1). \tag{7.4.11}$$

If $\beta < 0$, then putting $\alpha = 1 + \beta$ converts eqn. (7.4.11) to the exponential smoothing formula (7.3.3). Thus, the Box–Jenkins method provides a rationale for exponential smoothing in terms of an assumed statistical model. The obvious benefit of this is that the adequacy of this model can be assessed, and the parameter β estimated, using the methods described in Chapter 6.

This connection between exponential forecasting and a particular ARIMA process explains why the best forecasts for the wool price ratio were obtained using exponential smoothing with $\alpha = 1$. For, in Chapter 6 we showed that the best ARIMA model for these data is a simple random walk. This corresponds to the ARIMA(0,1,1) model (7.4.10) but with $\beta = 0$, i.e. $\alpha = 1 + \beta = 1$.

A minor refinement concerns a non-zero mean value for the first difference of the wool price ratio. In Example 6.4 we estimated this mean value to be $\hat{\mu} = 0.001\,556$ and noted that this value was not quite significantly different from zero. Now, a genuinely non-zero mean μ in an ARIMA($p,1,q$) model implies a linear trend in the undifferenced data, which should be incorporated into the forecasting equation. Specifically, if $Y_t(k)$ is the k-step ahead forecast assuming a zero mean, then the corresponding forecast when $\mu \neq 0$ is $Y_t(k) + k\mu$. For the log-wool-price-ratio, the estimate $\hat{\mu}$ is sufficiently close to zero that its inclusion has only a small effect on the forecasting performance. For one-step ahead forecasts, the rmse value decreases from 0.0133 to 0.0131, whilst for the ten-step ahead forecasts it decreases from 0.0574 to 0.0526.

When an ARIMA model-fitting exercise leads to a relatively large value of $\hat{\mu}$, the investigator has to question whether this is indeed indicative of a sustained linear trend, or of a stochastic trend which happens to have been observed in an increasing phase. Often, this issue can be resolved by inspection of the complete historical record, or by knowledge of the practical context. Note that when we fit an ARIMA($p,1,q$) model with a non-zero mean, μ is approximately the average of the first differences of the observations y_t. This is simply the difference between the last and first observation, so quite radically different historical records could lead to very similar values of $\hat{\mu}$.

The Box–Jenkins methodology allows the investigator to tailor a forecasting scheme to a particular series, based on an objective assessment of its historical behaviour. Whilst this does not guarantee success, for the reasons outlined at the beginning of this chapter, it nevertheless provides a sounder foundation than the more *ad hoc* methods described in Sections 7.2 and 7.3.

Another interesting consequence of a model-based methodology is that it allows for an assessment of the statistical uncertainty inherent in the forecasts it produces. The key to this is eqn. (7.4.8),

$$Y_{t+k} - Y_t(k) = \sum_{i=0}^{k-1} \theta_i Z_{t+k-i}.$$

Write $R_t(k) = Y_{t+k} - Y_t(k)$, the k-step ahead forecasting error at time t. Then,

$$E[R_t(k)] = \sum_{i=0}^{k-1} \theta_i E[Z_{t+k-i}] = 0,$$

and

$$V(k) = \text{var}\{R_t(k)\} = \sum_{i=0}^{k-1} \theta_i^2 \, \text{var}(Z_{t+k-i}) = \sigma^2 \sum_{i=0}^{k-1} \theta_i^2. \qquad (7.4.12)$$

Substituting the estimated values $\hat{\sigma}^2$ and $\hat{\theta}_i$ in eqn. (7.4.2) leads to an estimated forecasting variance, $\hat{V}(k)$ say, and tolerance intervals for the forecasts $Y_t(k)$ can be set by assuming that $R_t(k)$ is approximately Normally distributed with mean zero and variance $\hat{V}(k)$. Then, an approximate $100(1 - \alpha)\%$ tolerance interval for the k-step ahead forecast would be

$$Y_t(k) \pm c_\alpha \surd\{\hat{V}(k)\}, \qquad (7.4.13)$$

where c_α is the two-sided α-critical vlaue of the standard Normal distribution. For example, for a 90% tolerance interval we use $c_{0.1} = 1.645$.

Note that (7.4.13) ignores the statistical variation in the model parameter estimates and, probably more importantly, the increased uncertainty due to the imperfect relationship between the fitted model and the truth.

Example 7.6. Statistical variation in forecasts of the log-wool-price-ratio
The model fitted to the log-wool-price-ratio in Example 6.4 is a simple random walk,

$$Y_t = Y_{t-1} + Z_t,$$

with the variance of Z_t estimated as $\hat{\sigma}^2 = 0.000\ 190$. The k-step ahead forecast is just Y_t. To obtain an expression for the forecasting variance $V(k)$, we first need to express Y_t as a linear combination of past and current Z_t. To do this, we write the model in operator notation as

$$(1 - B)Y_t = Z_t$$

or

$$Y_t = (1 - B)^{-1}Z_t,$$

and apply a formal power series expansion of $(1 - B)^{-1}$ to give

$$Y_t = \left(\sum_{i=0}^{\infty} B^i\right)Z_t$$

$$= \sum_{i=0}^{\infty} Z_{t-i},$$

and similarly,

$$Y_{t+k} = \sum_{i=0}^{\infty} Z_{t+k-i}.$$

This last expression is of the form (7.4.8), with all $\theta_i = 1$, and (7.4.12) accordingly gives

$$V(k) = k\sigma^2.$$

Substitution of this expression for $V(k)$ into (7.4.13) then gives the form

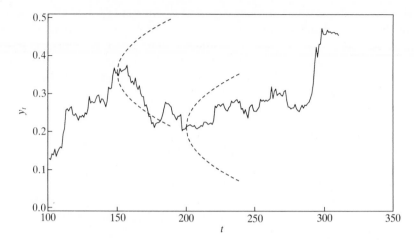

Fig. 7.1. Forecasts for the log-wool-price-ratio. The bowl-shaped masks represent 90% tolerance intervals for k-step ahead forecasts from $t = 150$ and from $t = 200$.

of the 90% tolerance interval for the k-step ahead forecast as

$$Y_t \pm 1.645\hat{\sigma}\sqrt{k}. = Y_t \pm 0.0227\sqrt{k}.$$

Figure 7.1 illustrates an application of this result to the log-wool-price ratio. Superimposed on the plot of the data are two bowl-shaped 'masks' which indicate the 90% tolerance limits on the k-step ahead forecasts for k up to 40, from $t = 150$ and from $t = 200$. From $t = 150$, the series is in a

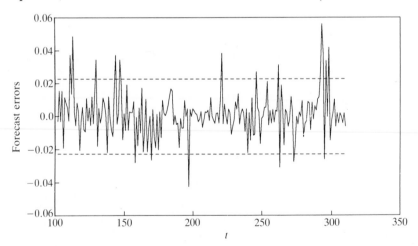

Fig. 7.2. One-step ahead forecast errors, $Y_{t+1} - Y_t(1)$, for the log-wool-price-ratio. The dashed horizontal lines represent pointwise 90% tolerance limits for these errors.

sharply decreasing phase and the observed values eventually fall below the lower limit, whereas from $t = 200$ the series is in a more stable phase and the observations remain well within the limits.

Figure 7.2 illustrates a slightly different application to the same data. In this case, one-step forecast errors, $Y_{t+1} - Y_t(1)$, are plotted along with the 90% tolerance limits for these errors. As we would expect, a small proportion of the errors falls outside the limits: in fact, 19 out of the 209 do so, in good agreement with the theoretical 10% of forecasts which we would expect to lie outside the 90% limits.

7.5 Further reading

Much of the literature on forecasting is motivated by business applications (Whittle 1984 is a notable exception), although the associated methodology tends to be context-free. Gilchrist (1976) and Abraham and Ledolter (1983) are two general introductions to statistical forecasting in which the Box–Jenkins approach and other approaches based on linear models feature prominently. Jenkins (1979) gives a series of case-studies using this approach. Harrison and Stevens (1976) and West et al. (1985) develop a quite different approach using a Bayesian interpretation of the Kalman filter (see Section 3.7). Makridakis et al. (1984) present the results of a 'forecasting competition' in which the contributors apply competing methodologies to a large number of economic time series.

8
Elements of bivariate time-series analysis

8.1 Introduction

In this final chapter we give a brief introduction to the analysis of data in the form of a pair of possibly inter-dependent time series. Our notation for this will be $\{(x_t, y_t) : t = 1, \ldots, n\}$. For motivation, consider Example 1.4 involving monthly numbers of male and female deaths in the UK due to bronchitis, emphysema, and asthma. Both series show an obvious seasonal pattern. Might there also be a more subtle form of inter-dependence, for example arising out of a common response in both series to unusually harsh or mild weather conditions? How should we analyse the data to see if this is indeed the case?

Our theoretical model will be a pair of random processes, which we write as $\{(X_t, Y_t)\}$. In order to analyse the inter-dependence between $\{X_t\}$ and $\{Y_t\}$ we shall assume that both processes are stationary. It follows that the observed series $\{X_t\}$ and $\{Y_t\}$ must be assumed to be stationary, possibly after trend removal by one or other of the methods described in Chapter 2.

In Section 8.2 we define the *cross-covariance* and *cross-correlation* functions of a stationary bivariate process $\{(X_t, Y_t)\}$. In Section 8.3 we discuss estimation of the cross-correlation function from data via the *cross-correlogram*. We also illustrate the danger inherent in naïve interpretation of the cross-correlogram and show how this danger can be alleviated by *pre-whitening* the data. In Section 8.4 we describe the frequency-domain analogue of the cross-covariance function, the *cross-spectrum*.

8.2 The cross-covariance and cross-correlation functions

The cross-covariance function of a bivariate stationary random process $\{(X_t, Y_t)\}$ is the set of numbers $\gamma_{xy}(k)$ defined, for each integer k, by

$$\gamma_{xy}(k) = \mathrm{cov}\{X_t, Y_{t-k}\}.$$

Note that $\gamma_{xy}(k)$ does not depend on t because $\{X_t\}$ and $\{Y_t\}$ are both stationary processes, and that $\gamma_{xx}(k)$ and $\gamma_{yy}(k)$ are the autocovariance functions of $\{X_t\}$ and $\{Y_t\}$ respectively. Also

$$
\begin{aligned}
\gamma_{xy}(-k) &= \text{cov}\{X_t, Y_{t+k}\} \\
&= \text{cov}\{Y_{t+k}, X_t\} \\
&= \text{cov}\{Y_t, X_{t-k}\} = \gamma_{yx}(k).
\end{aligned}
$$

We therefore need consider only one of $\gamma_{xy}(k)$ and $\gamma_{yx}(k)$, but we do need to consider both positive and negative integer values of k.

The cross-correlation function of $\{X_t\}$ and $\{Y_t\}$ is

$$
\rho_{xy}(k) = \gamma_{xy}(k)/\sqrt{\{\gamma_{xx}(0)\gamma_{yy}(0)\}}.
$$

A specific example may be illuminating. Suppose that $\{X_t\}$ is a first-order autoregressive process,

$$
X_t = \alpha X_{t-1} + Z_t, \tag{8.2.1}
$$

where $-1 < \alpha < 1$ and $\{Z_t\}$ is a white noise process with variance σ^2. Suppose also that $\{Y_t\}$ is related to $\{X_t\}$ by a linear regression of the form

$$
Y_t = \beta X_{t-l} + W_t, \tag{8.2.2}
$$

where $\{W_t\}$ is a white noise process with variance τ^2, independent of both $\{Z_t\}$ and $\{X_t\}$.

Intuitively, we might expect the cross-correlation $\rho_{xy}(k)$ to be maximized at $k = -l$, to reflect the time lag in the regression relationship between $\{Y_t\}$ and $\{X_t\}$.

Now, we already know from results in Section 3.4 that the auto-covariance function of $\{X_t\}$ is

$$
\gamma_{xx}(k) = \sigma^2 \alpha^k / (1 - \alpha^2).
$$

Also, the autocovariance function of $\{Y_t\}$ is

$$
\begin{aligned}
\gamma_{yy}(k) &= E[Y_t Y_{t-k}] \\
&= E[(\beta X_{t-l} + W_t)(\beta X_{t-k-l} + W_{t-k})] \\
&= \beta^2 E[X_{t-l} X_{t-k-l}] = \beta^2 \gamma_{xx}(k),
\end{aligned}
$$

provided $k \neq 0$. For $k = 0$,

$$
\begin{aligned}
\gamma_{yy}(0) &= E[Y^2] \\
&= \beta^2 E[X_{t-l}^2] + E[W_t^2] \\
&= \beta^2 \sigma^2 / (1 - \alpha^2) + \tau^2 = \beta^2 \gamma_{xx}(0) + \tau^2.
\end{aligned}
$$

Finally, the cross-covariance function is

$$\gamma_{xy}(k) = E[X_t Y_{t-k}]$$
$$= E[X_t(\beta X_{t-k-l} + W_{t-k})]$$
$$= \beta \gamma_{xx}(k+l),$$

so the cross-correlation function is

$$\rho_{xy}(k) = \beta \gamma_{xx}(k+l)/\sqrt{\{\gamma_{xx}(0)(\beta^2\gamma_{xx}(0) + \tau^2)\}}$$
$$= \gamma_{xx}(k+l)/\sqrt{\{\gamma_{xx}(0)(\gamma_{xx}(0) + \tau^2/\beta^2)\}}$$
$$= \alpha^{k+l}/\sqrt{\{1 + \tau^2(1 - \alpha^2)/(\beta^2\sigma^2)\}}. \qquad (8.2.3)$$

Clearly, the cross-correlation is maximized at $k = -l$, as anticipated, and decays exponentially on either side of $k = -l$, with alternating sign if $\alpha < 0$. Also, the maximum value $\rho_{xy}(-l)$ depends on the values of the parameters in the model in a sensible way. Firstly, for fixed α, σ^2 and τ^2, $\rho_{xy}(-l)$ increases with the absolute value of the regression coefficient β. Secondly, for fixed β, $\rho_{xy}(-l)$ is a function of the ratio between $\text{var}(W_t) = \tau^2$ and $\text{var}(X_t) = \sigma^2/(1 - \alpha^2)$, and decreases as this ratio increases.

8.3 Estimating the cross-correlation function

A natural estimator for the cross-correlation function is the *cross-correlogram*, defined as follows. For a bivariate series $\{(x_t, y_t): t = 1, \ldots, n\}$, the *cross-correlogram* is the set of numbers

$$r_{xy}(k) = g_{xy}(k)/\sqrt{\{g_{xx}(0)g_{yy}(0)\}}$$

where

$$g_{xx}(0) = n^{-1}\sum_{t=1}^{n}(x_t - \bar{x})^2, \qquad g_{yy}(0) = n^{-1}\sum_{t=1}^{n}(y_t - \bar{y})^2,$$

$$\bar{x} = n^{-1}\sum_{t=1}^{n}x_t, \qquad \bar{y} = n^{-1}\sum_{t=1}^{n}y_t,$$

and

$$g_{xy}(k) = \begin{cases} n^{-1}\displaystyle\sum_{t=k+1}^{n}(x_t - \bar{x})(y_{t-k} - \bar{y}), & k \geq 0 \\ n^{-1}\displaystyle\sum_{t=1}^{n+k}(x_t - \bar{x})(y_{t-k} - \bar{y}), & k < 0. \end{cases}$$

Example 8.1. Cross-correlogram of simulated data
Figure 8.1 shows a simulation of the model defined by eqns (8.2.1) and (8.2.2), with $\alpha = 0.5$, $\beta = 0.9$, $\sigma^2 = \tau^2 = 1$, and $l = 2$. Note that we have

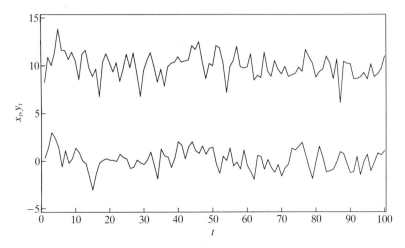

Fig. 8.1. A simulated realization of a bivariate stationary random process.

added 10 to each plotted value of y_t to keep the two traces separate. Note also how the rise-and-fall pattern in the x_t-trace is reflected in the y_t-trace, but with a time lag of two units. Figure 8.2 shows the cross-correlogram of these data together with the theoretical form of the cross-correlation function. The two are in close agreement.

In the simulated example above, the cross-correlogram gave a good estimate of the underlying cross-correlation function. This is not always

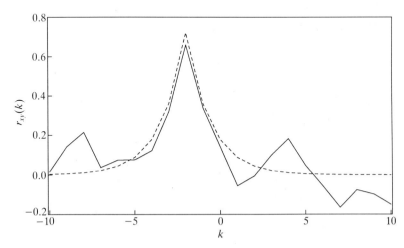

Fig. 8.2. Cross-correlation structure of the data from Fig. 8.1. ———, The cross-correlogram of the data; $---$, The cross-correlation function of the model used to simulate the data.

the case. To see why, suppose that data $\{(x_t, y_t): t = 1, \ldots, n\}$ are generated by a pair of independent AR(1) process with a common autoregressive parameter α. Because the underlying processes are independent, $\rho_{xy}(k) = 0$, for all k, irrespective of the value of α. However, if for example α is very close to -1, the observed series $\{x_t\}$ and $\{y_t\}$ will each tend to contain long sequences of alternating positive and negative values. Consequently, the sample cross-correlation $r_{xy}(0)$ may be quite large in absolute value, and positive or negative according to whether, by chance, the two alternating sequences are in-phase or out-of-phase. Essentially the same phenomenon arises when α is close to $+1$: in that case, each of $\{x_t\}$ and $\{y_t\}$ tend to contain long subsequences of all positive or all negative values which again may be in phase or out of phase by chance.

The argument in the previous paragraph suggests that, whilst the sample cross-correlogram of a pair of independently generated time series $\{x_t\}$ and $\{y_t\}$ may well have expectation close to zero, its sampling distribution and in particular its variance will depend on the marginal properties of the two series. Figure 8.3 summarizes the results of a small-scale simulation experiment in which we generated pairs of series $\{x_t : t = 1, \ldots, 100\}$ and $\{y_t : t = 1, \ldots, 100\}$ from an AR(1) process with autoregressive parameter α. For each value of α, we generated 100 pairs

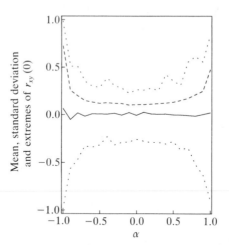

Fig. 8.3. The results of a small-scale simulation experiment concerning the sampling distribution of the zero-lag cross-correlation, $r_{xy}(0)$, when the data are two independent series of $n = 100$ observations, each generated by an AR(1) process with autoregressive parameter α. The diagram shows, as functions of α, the following summary statistics for $r_{xy}(0)$ over 100 replicates of the experiment. ——, Sample mean; – – –, sample standard deviation; - - - -, lower and upper extremes.

of independent series and calculated the zero-lag cross-correlation $r_{xy}(0)$ for each pair. We see that, whilst the average value of $r_{xy}(0)$ is close to zero in all cases, both the sample standard deviation and range of $r_{xy}(0)$ increase with α.

Note that in Fig. 8.3 we have included results for the cases $\alpha = 1$ and $\alpha = -1$. Because these correspond to non-stationary processes $\{X_t\}$ and $\{Y_t\}$, the theoretical cross-correlation $\rho_{xy}(0)$ is undefined. We can, of course, still calculate $r_{xy}(0)$, but its interpretation is meaningless. Another example would be one in which the two series are independent, but with a common trend. Then, values of $r_{xy}(0)$ quite close to unity might easily occur, but this apparent correlation between the two series would be entirely spurious.

More formally, for data $\{(x_t, y_t) : t = 1, \ldots, n\}$ generated from a stationary, bivariate process $\{(X_t, Y_t)\}$, $r_{xy}(k)$ is an approximately unbiased estimator for $\rho_{xy}(k)$, and its variance is

$$\text{var}\{r_{xy}(k)\} = n^{-1} \sum_{l=-\infty}^{\infty} \{\rho_{xx}(l)\rho_{yy}(l) + \rho_{xy}(l+k)\rho_{xy}(l-k)\}.$$

In particular, if $\{X_t\}$ and $\{Y_t\}$ are independent processes, so that $\rho_{xy}(k) = 0$ for all k, then

$$\text{var}\{r_{xy}(k)\} = n^{-1} \sum_{l=-\infty}^{\infty} \rho_{xx}(l)\rho_{yy}(l). \tag{8.3.1}$$

In our simulation study of $r_{xy}(0)$, we used an autoregressive model for which

$$\rho_{xx}(l) = \rho_{yy}(l) = \alpha^l.$$

Then, (8.3.1) gives

$$\text{var}\{r_{xy}(k)\} = n^{-1}(1 + \alpha^2)/(1 - \alpha^2), \tag{8.3.2}$$

showing in particular that the variance of $r_{xy}(0)$ increases as $|\alpha|$ increases from 0 to 1, gently at first but more steeply as $|\alpha|$ approaches 1. The empirical results in Fig. 8.3 are in good qualitative agreement with this. Note, however, that the approximation (8.3.2) becomes poor as $|\alpha|$ approaches 1.

Motivated by the above discussion, we might consider whether interpretation of the cross-correlogram would not be easier if we first transformed the series $\{x_t\}$ and $\{y_t\}$ so that each resembled a realization of white noise. We can then assess the significance of the cross-correlogram using the approximate formula,

$$\text{var}\{r_{xy}(k)\} \simeq n^{-1}.$$

In interpreting the cross-correlogram using the above formula, the same qualification is in order as for the univariate case: successive $r_{xy}(k)$ are themselves correlated, and the overall pattern of the cross-correlogram is usually of more interest than the precise numerical values of individual $r_{xy}(k)$.

Transformation of observed series to resemble white noise prior to evaluation of their inter-dependence is called *pre-whitening*. The usual way to pre-whiten is to apply a linear filter, in the form of either moving average trend elimination or ARIMA model fitting. In the former case, we subtract from the observed series a moving average, thereby eliminating trends in either series. In the latter case, we extract from each series the embedded residual series based on the fitted ARIMA model.

Note that pre-whitening does affect the interpretation to be placed on any significant cross-correlation structure. Quite generally, if we define

$$U_t = \sum_{i=0}^{\infty} a_i X_{t-i}$$

and

$$V_t = \sum_{j=0}^{\infty} b_j Y_{t-j},$$

then provided the coefficients a_i and b_j are such as to preserve stationarity, the cross-covariance functions of $\{(X_t, Y_t)\}$ and $\{(U_t, V_t)\}$ are related by

$$\gamma_{uv}(k) = \sum_{i=0}^{\infty} \sum_{j=0}^{\infty} a_i b_j \gamma_{xy}(k + i - j).$$

Example 8.2. Monthly UK deaths from bronchitis, emphysema, and asthma

Figure 8.4 shows the cross-correlogram for the two series from Example 1.4 relating to male and female deaths. The pattern in the cross-correlogram is a direct reflection of the strong seasonal pattern in the data. A more interesting exercise is to eliminate the seasonal trend from the data and examine the two trend-free series for possible dependence. In Example 2.2, we used a three-point moving average to estimate the seasonal trend. Subtracting this trend estimate from each series leaves the two residual series plotted in Fig. 8.5, where we have added 600 to each of the male residuals to keep the two traces apart. Figure 8.6 shows the correlogram of these two residual series. The pre-whitening has not been entirely successful, because in both cases the autocorrelation at lag one is significantly negative; this is in part a consequence of the moving average method of trend removal (cf. Section 2.6). Figure 8.7 shows the cross-correlogram of the two residual series. The plot also includes a pair

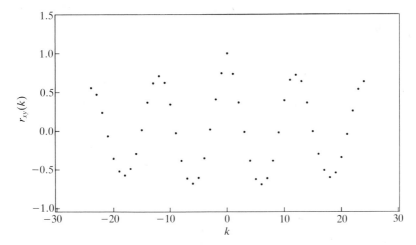

Fig. 8.4. The cross-correlogram of male and female monthly returns of deaths in the United Kingdom attributed to bronchitis, emphysema, and asthma.

of horizontal lines for assessing the significance of the estimated cross-correlations. These are derived as follows.

Figure 8.6 suggests that both residual series have a negative autocorrelation at lag one and that higher-order autocorrelations are approximately zero. Assume that the theoretical autocorrelations are exactly

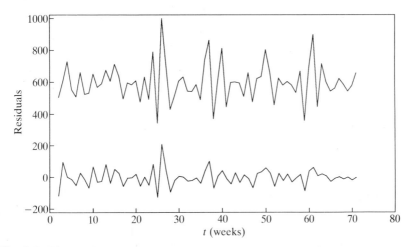

Fig. 8.5. The residual series obtained by subtracting three-point moving averages from the monthly returns of deaths in the United Kingdom attributed to bronchitis, emphysema, and asthma. Note that 600 has been added to each male residual.

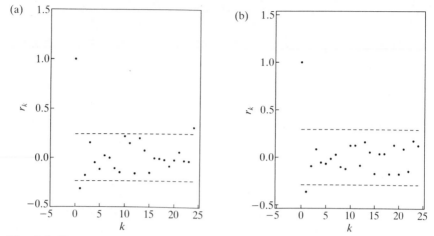

Fig. 8.6. The correlograms of the two residual series shown in Fig. 8.5. The dashed horizontal lines correspond to the limits $\pm 2/\sqrt{n}$. (a) Males. (b) Females.

zero for all lags greater than one. Then, eqn. (8.3.1) becomes,

$$\operatorname{var}\{r_{xy}(k)\} \simeq n^{-1}\{1 + 2\rho_{xx}(1)\rho_{yy}(1)\}. \qquad (8.3.3)$$

Now, substitute into (8.3.3) the sample autocorrelations, $r_{xx}(1) = -0.315$ and $r_{yy}(1) = -0.363$, to give

$$\operatorname{var}\{r_{xy}(k)\} \simeq 1.229 n^{-1}.$$

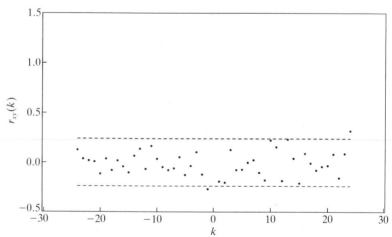

Fig. 8.7. The cross-correlogram of the two residual series shown in Fig. 8.5. The dashed horizontal lines correspond to approximate pointwise 95% tolerance limits under the assumption that the underlying cross-correlation function is zero at all lags.

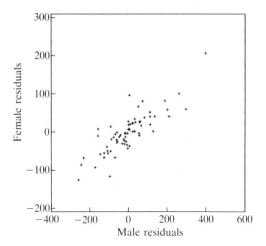

Fig. 8.8. A scatter plot of female residuals against contemporaneous male residuals.

Finally, construct approximate pointwise 95% tolerance limits to assess departure from zero cross-correlations as $\pm 2.217/\sqrt{n}$, these representing plus and minus two standard errors under the assumed forms for the autocorrelation structure of the two residual series. Note that these limits are only slightly wider than the $\pm 2/\sqrt{n}$ limits which would be appropriate if both residual series were realizations of white noise.

The clear message from Fig. 8.7 is that there is a strong positive cross-correlation at lag zero, but no other noteworthy structure; although the sample cross-correlations at lags -1 and 24 lie just outside the $\pm 2.217/\sqrt{n}$ limits, this is unremarkable in a sequence of 48 sample cross-correlations. To explore the dependence further, Fig. 8.8 shows a scatter plot of the values of one residual series against the contemporaneous values of the other. The association appears to be approximately linear, so the value of the cross-correlation coefficient is a reasonable summary. Incidentally, if we remove the extreme point in the top right- hand corner of Fig. 8.3, corresponding to January 1976, when both male and female monthly deaths reached their highest numbers during the six-year period covered by the data, the cross-correlation is reduced from 0.83 to 0.64, which is still highly significant. We therefore conclude that contemporaneous fluctuations about the seasonal trends in male and female deaths are positively correlated, suggesting that these fluctuations are at least in part the result of a common external factor, plausibly the vagaries of the British climate. Lack of such correlation would have suggested that the deviations were the result of random variation in the susceptibilities of the two populations at risk.

8.4   The spectrum of a bivariate process

By analogy with the univariate case, we can define the *cross-spectrum*, $f_{xy}(\omega)$, of a bivariate stationary process $\{(X_t, Y_t)\}$ as the discrete Fourier transform of its autocovariance function $\gamma_{xy}(k)$,

$$f_{xy}(\omega) = \sum_{k=-\infty}^{\infty} \gamma_{xy}(k)e^{-ik\omega}. \qquad (8.4.1)$$

Because in general $\gamma_{xy}(k) \neq \gamma_{xy}(-k)$, the cross-spectrum is in general a complex-valued function. This makes it difficult to interpret. However, it is possible to provide a tangible interpretation by considering its decomposition into real and imaginary parts.

Firstly, write $f_{xy}(\omega)$ as $c_{xy}(\omega) - iq_{xy}(\omega)$ (the minus sign is conventional in this context) and note from (8.4.1) that

$$c_{xy}(\omega) = \sum_{k=-\infty}^{\infty} \gamma_{xy}(k)\cos(k\omega)$$

$$= \gamma_{xy}(0) + \sum_{k=1}^{\infty} \{\gamma_{xy}(k) + \gamma_{xy}(-k)\}\cos(k\omega)$$

and

$$q_{xy}(\omega) = \sum_{k=1}^{\infty} \{\gamma_{xy}(k) - \gamma_{xy}(-k)\}\sin(k\omega).$$

The functions $c_{xy}(\omega)$ and $q_{xy}(\omega)$ are called the *co-spectrum* and *quadrature spectrum*, respectively. Now, consider the polar representation of the cross-spectrum,

$$f_{xy}(\omega) = a_{xy}(\omega)\exp\{i\phi_{xy}(\omega)\}.$$

Then,

$$a_{xy}(\omega) = |f_{xy}(\omega)| = \sqrt{\{c_{xy}(\omega)^2 + q_{xy}(\omega)^2\}}$$

is called the *cross-amplitude* spectrum and

$$\phi_{xy}(\omega) = \tan^{-1}\{-q_{xy}(\omega)/c_{xy}(\omega)\}$$

the *phase spectrum*. We shall consider in turn the physical interpretation of the cross-amplitude spectrum and of the phase spectrum.

The cross-amplitude spectrum represents a form of covariance between the frequency components of $\{X_t\}$ and $\{Y_t\}$ at frequency ω in the same way that the marginal spectra $f_{xx}(\omega)$ and $f_{yy}(\omega)$ represent the variances of the corresponding frequency components of $\{X_t\}$ and of $\{Y_t\}$. Thus, the quantity

$$b_{xy}(\omega) = a_{xy}(\omega)/\sqrt{\{f_{xx}(\omega)f_{yy}(\omega)\}},$$

called the *coherency* between $\{X_t\}$ and $\{Y_t\}$ at frequency ω, represents the correlation between the corresponding frequency components of $\{X_t\}$ and $\{Y_t\}$.

Note that the coherency is necessarily non-negative; in fact, $0 \leqslant b_{xy}(\omega) \leqslant 1$. This brings us to the interpretation of the phase spectrum. Essentially, the coherency is non-negative because it measures the correlation between *aligned* frequency components. For example, if $\{X_t\}$ shows a strong pattern of alternating positive and negative values and $Y_t \simeq -X_t$, then the processes $\{X_t\}$ and $\{Y_t\}$ will be strongly coherent but out of phase, the latter property being measured by the *phase spectrum*.

An example may help to clarify the discussion. Suppose that $\{X_t\}$ is a stationary process with autocovariance function $\gamma_{xx}(k)$ and spectrum $f_{xx}(\omega) = \sum_{k=-\infty}^{\infty} \gamma_{xx}(k)e^{-ik\omega}$. Let $\{Y_t\}$ be defined by

$$Y_t = -X_t + Z_t,$$

where $\{Z_t\}$ is a white noise process with variance τ^2, independent of $\{X_t\}$. The spectrum of $\{Y_t\}$ is

$$f_{yy}(\omega) = f_{xx}(\omega) + \tau^2,$$

and the cross-covariance function and cross-spectrum are

$$\gamma_{xy}(k) = -\gamma_{xx}(k)$$

and

$$f_{xy}(\omega) = -f_{xx}(\omega). \tag{8.4.2}$$

If we now express (8.4.2) in terms of the cross-amplitude and phase spectra, we have

$$-f_{xx}(\omega) = a_{xy}(\omega) \exp\{i\phi_{xy}(\omega)\}.$$

However, we know that $f_{xx}(\omega)$ is a real-valued function, and that $a_{xy}(\omega)$ is non-negative by definition, from which it follows that

$$a_{xy}(\omega) = f_{xx}(\omega)$$

and

$$\phi_{xy}(\omega) = \pi, \tag{8.4.3}$$

for all ω, since $\exp(i\pi) = -1$. Finally, the coherency between $\{X_t\}$ and $\{Y_t\}$ is

$$b_{xy}(\omega) = f_{xx}(\omega)/\sqrt{[f_{xx}(\omega)\{f_{xx}(\omega) + \tau^2\}]}$$
$$= \{1 + \tau^2/f_{xx}(\omega)\}^{-1/2}. \tag{8.4.4}$$

Note how the forms of the phase spectrum (8.4.3) and the coherency (8.4.4) make intuitive sense. The constant value of π radians for the phase spectrum corresponds to the simple 'change-of-sign' relationship between $\{X_t\}$ and $\{Y_t\}$. Also, the coherency is small if τ^2 is large and vice versa, according as the relationship between X_t and Y_t is or is not swamped by the random variation in the white noise sequence $\{Z_t\}$. Note also that we do need the phase spectrum to be able to take values over the full circular range of its argument, for example $-\pi < \omega < \pi$, which we achieve by insisting that $\phi_{xy}(\omega)$ have the opposite sign to $q_{xy}(\omega)$.

A second example, in which the phase spectrum does not simply assume a constant value for all ω, reinforces its physical interpretation. suppose that $\{X_t\}$ and $\{Y_t\}$ are related by

$$Y_t = \beta X_{t-l} + Z_t,$$

where again $\{Z_t\}$ is a white noise sequence with variance τ^2, independent of $\{X_t\}$. In Section 8.2 we showed that for this bivariate process,

$$\gamma_{xy}(k) = \beta \gamma_{xx}(k + l),$$
$$\gamma_{yy}(k) = \beta^2 \gamma_{xx}(k)$$

for $k \neq 0$ and

$$\gamma_{yy}(0) = \beta^2 \gamma_{xx}(0) + \tau^2.$$

Thus, the cross-spectrum is

$$f_{xy}(k) = \sum_{k=-\infty}^{\infty} \gamma_{xy}(k)e^{-ik\omega}$$

$$= \beta \sum_{k=-\infty}^{\infty} \gamma_{xx}(k + l)e^{-ik\omega}$$

$$= \beta e^{il\omega} \sum_{k=-\infty}^{\infty} \gamma_{xx}(k + l)e^{-i(k+l)\omega}$$

$$= \beta f_{xx}(\omega)e^{il\omega}. \tag{8.4.5}$$

Also,

$$f_{yy}(\omega) = \beta^2 f_{xx}(\omega) + \tau^2.$$

The cross-amplitude spectrum is therefore

$$a_{xy}(\omega) = |\beta| f_{xx}(\omega),$$

and the coherency is

$$b_{xy}(\omega) = [1 + \tau^2/\{\beta^2 f_{xx}(\omega)\}]^{-1/2}, \tag{8.4.6}$$

where $f_{xx}(\omega)$ is the spectrum of an AR(1) process,

$$f_{xx}(\omega) = \sigma^2/(1 - 2\alpha \cos \omega + \alpha^2). \tag{8.4.7}$$

If $\beta > 0$, it follows from (8.4.5) that the phase spectrum is

$$\phi_{xy}(\omega) = l\omega, \tag{8.4.8}$$

whereas if $\beta < 0$,

$$\phi_{xy}(\omega) = \pi + l\omega.$$

The important point to note is that *the slope of the phase spectrum corresponds to the delay between $\{X_t\}$ and $\{Y_t\}$.*

This last observation provides the key to the following general interpretation of coherency and phase spectra. Each if $\{X_t\}$ and $\{Y_t\}$ can be represented as a superposition of stochastic components at frequencies in the range 0 to π; the spectrum of $\{X_t\}$ represents a partitioning of the variability in $\{X_t\}$ into its different frequency components, and similarly for the spectrum of $\{Y_t\}$; the coherency at frequency ω represents the strength of the correlation between the corresponding frequency components of $\{X_t\}$ and $\{Y_t\}$; finally, the slope of the phase spectrum at frequency ω represents the time delay between these frequency components.

8.5 Estimation of the cross-spectrum

An obvious estimator for the cross-spectrum is obtained by substituting the sample cross-covariances $g_{xy}(k)$ for the theoretical cross-covariances $\gamma_{xy}(k)$ in eqn. (8.4.1) and truncating the infinite summation. The resulting estimate is the *cross-periodogram*,

$$I_{xy}(\omega) = \sum_{k=-(n-1)}^{n-1} g_{xy}(k)e^{-ik\omega}, \qquad (8.5.1)$$

where n denotes the length of the observed series $\{x_t\}$ and $\{y_t\}$. An alternative expression, which shows how the cross-periodogram can be evaluated using a fast Fourier transform, is

$$I_{xy}(\omega) = n^{-1}J_x(\omega)J_y^*(\omega), \qquad (8.5.2)$$

where J^* is the complex conjugate of J, and

$$J_x(\omega) = \sum_{t=1}^{n} x_t e^{-i\omega t}. \qquad (8.5.3)$$

A third expression, obtained by dividing eqn. (8.5.3) into real and imaginary parts, is

$$I_{xy}(\omega) = n^{-1}\{A_x(\omega)A_y(\omega) + B_x(\omega)B_y(\omega)\}$$
$$\qquad - n^{-1}i\{B_x(\omega)A_y(\omega) - A_x(\omega)B_y(\omega)\}$$
$$= \tilde{c}(\omega) - i\tilde{q}(\omega), \qquad (8.5.4)$$

where $\tilde{c}(\omega)$ and $\tilde{q}(\omega)$ are estimates of the co-spectrum and quadrature spectrum, and

$$A_x(\omega) = \sum_{t=1}^{n} x_t \cos(t\omega),$$

$$B_x(\omega) = \sum_{t=1}^{n} x_t \sin(t\omega),$$

with analogous expressions for $A_y(\omega)$, $B_y(\omega)$.

In all these formulae, ω is understood to be of the form $\omega = 2\pi j/n$, with j a positive integer less than or equal to $n/2$.

The estimates $\bar{c}(\omega)$ and $\bar{q}(\omega)$ in eqn. (8.5.4) suffer exactly the same defects as the ordinary periodogram of a single series, and will usually benefit from smoothing along the lines described in Chapter 4. In particular, smoothing by a simple moving average of order $2p + 1$ leads to estimates

$$\hat{c}(\omega_j) = (2p + 1)^{-1} \sum_{k=-p}^{p} \bar{c}(\omega_{j+k}),$$

$$\hat{q}(\omega_j) = (2p + 1)^{-1} \sum_{k=-p}^{p} \bar{q}(\omega_{j+k}),$$

where $\omega_j = 2\pi j/n$.

Having obtained satisfactorily smooth estimates $\hat{c}(\omega)$ and $\hat{q}(\omega)$, we convert these to estimates of the cross-amplitude and phase,

$$\hat{a}_{xy}(\omega) = \{\hat{c}(\omega)^2 + \hat{q}(\omega)^2\}, \tag{8.5.3}$$

$$\hat{d}_{xy}(\omega) = \tan^{-1}\{-\hat{q}(\omega)/\hat{c}(\omega)\}, \tag{8.5.6}$$

and of the coherency,

$$\hat{b}_{xy}(\omega) = \hat{a}_{xy}(\omega)/\sqrt{\{\hat{f}_{xx}(\omega)\hat{f}_{yy}(\omega)\}}, \tag{8.5.7}$$

where $\hat{f}_{xx}(\omega)$ and $\hat{f}_{yy}(\omega)$ are estimates of the two marginal spectra. The sampling variances of these smoothed estimates are

$$\text{var}\{\hat{a}_{xy}(\omega)\} = c_p\{a_{xy}(\omega)\}^2[1 + \{b_{xy}(\omega)\}^{-2}], \tag{8.5.8}$$

$$\text{var}\{\hat{\phi}_{xy}(\omega)\} = c_p[\{b_{xy}(\omega)^{-2}\} - 1], \tag{8.5.9}$$

and

$$\text{var}\{\hat{b}_{xy}(\omega)\} = c_p[1 - \{b_{xy}(\omega)\}^2]^2, \tag{8.5.10}$$

where $c_p = \{2(2p + 1)\}^{-1}$. For derivations of these results, see Jenkins and Watts (1968, Chapter 9). Note in particular that the sampling variances are large when the coherency is small, and when p is small. As in the univariate case the choice of p reflects a balance between the conflicting requirements of small bias, which requires small p, and small sampling variance, which requires large p.

A practical point concerning the plotting of the estimated phase spectrum (8.5.6) is that our convention about the inverse tangent, namely $-\pi \leqslant \tan^{-1}(x) \leqslant \pi$ for all x, is liable to produce discontinuities in $\hat{\phi}_{xy}(\omega)$. This is highly undesirable, because the slope of $\phi_{xy}(\omega)$ is the feature of most interest. A simple way round this is to plot $\hat{\phi}_{xy}(\omega)$ as a set of unconnected points, together with similar plots of $\hat{\phi}_{xy}(\omega) + 2\pi$ and $\hat{\phi}_{xy}(\omega) - 2\pi$. A continuous estimate of the phase spectrum can then be defined by joining points appropriately—whether literally in the plot or only in the mind's eye.

Example 8.3. Cross-spectral estimates for simulated data
Figure 8.9 shows the following for the simulated data of Fig. 8.1: the periodogram and smoothed periodogram of $\{x_t\}$ in Fig. 8.9a; the periodogram and smoothed periodogram of $\{y_t\}$ in Fig. 8.9b; the unsmoothed and smoothed estimates of the co-spectrum, $\bar{c}(\omega)$ and $\hat{c}(\omega)$, in Fig. 8.9c; the unsmoothed and smoothed estimates of the quadrature spectrum, $\bar{q}(\omega)$ and $\hat{q}(\omega)$, in Fig. 8.9d. In all four cases, smoothing was by a simple nine-point moving average. Figure 8.10 shows the estimates of the coherency and the phase spectrum derived from the smoothed estimates in Fig. 8.9, together with the theoretical coherency and phase spectrum. The theoretical forms are derived from eqns (8.4.6), (8.4.7), and (8.4.8) by substituting the parameter values $l = 2$, $\alpha = 0.5$, $\beta = 0.9$, and $\sigma^2 = \tau^2 = 1$, to give

$$b_{xy}(\omega) = (2.543 - 1.235 \cos \omega)^{-1/2}$$

and

$$\phi_{xy}(\omega) = 2\omega.$$

Example 8.4. Cross-spectral estimates for data on monthly UK deaths from bronchitis, emphysema and asthma
A cross-spectral analysis gives a different slant on the pattern of interdependence between the two residual series derived from these data by subtracting a three-pont moving average, as in Example 8.2. Figure 8.11 shows unsmoothed and smoothed estimates of the two spectra, the co-spectrum and the quadrature spectrum. In all four cases, a simple nine-point moving average was used for smoothing. Figure 8.12 shows the estimates of coherence and phase derived from the four smoothed estimates in Fig. 8.11. The dashed lines on Fig. 8.12 represent approximate 95% confidence limits. These were calculated as the estimated quantities plus and minus two standard deviations, which in turn were calculated from eqns (8.5.9) and (8.5.10), substituting the smoothed estimate of coherency for the unknown $b_{xy}(\omega)$.

We interpret the coherency and phase estimates as follows. Coherency

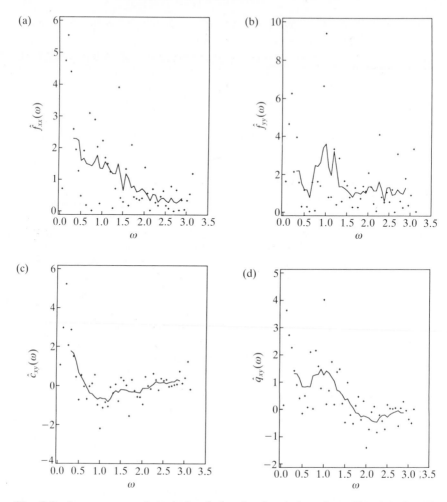

Fig. 8.9. A cross-spectral analysis of the simulated data from Fig. 8.1. In all cases, smoothing was carried out using a simple 9-point moving average. (a) The periodogram (+) and smoothed periodogram (——) of $\{x_t\}$. (b) The periodogram (+) and smoothed periodogram (——) of $\{y_t\}$. (c) The unsmoothed (+) and smoothed (——) estimates of the co-spectrum. (d) The unsmoothed (+) and smoothed (——) estimates of the quandrature spectrum.

is strongest at low frequencies, but remains substantial throughout the frequency range. The phase plot has a negative slope, approximately 0.1 in magnitude. At first sight, this might suggest that the $\{x_t\}$ series (males) *leads* the $\{y_t\}$ series (females) by approximately 0.1 months or three days. However, the relatively wide confidence limits on the estimated

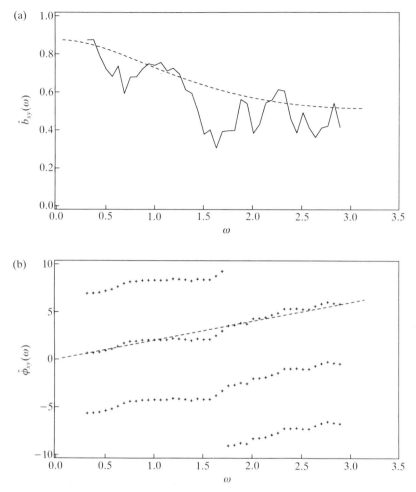

Fig. 8.10. The estimates of coherency and phase derived from the smoothed estimates shown in Fig. 8.9, and the corresponding theoretical functions for the model used to simulate the data. (a) Estimated (——) and theoretical (– – –) coherency. (b) Estimated (+) and theoretical (– – –) phase.

phase spectrum warn us against too narrow an interpretation. In particular, the data are compatible with an underlying zero phase spectrum, i.e. the two series are in alignment. The combination of a zero phase spectrum and strong coherence reinforces our earlier interpretation of these data, namely that fluctuations in the two residual series are largely the result of fluctuations in a common environmental factor which 'drives' both series.

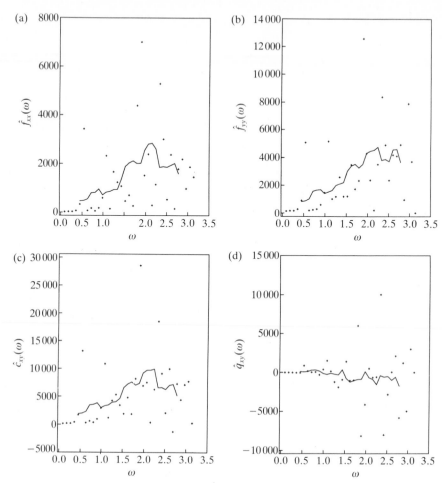

Fig. 8.11. A cross-spectral analysis of the two residual series shown in Fig. 8.5. In all cases, smoothing was carried out using a simple 9-point moving average. (a) The periodogram (+) and smoothed periodogram (——) of the male residuals. (b) The periodogram (+) and smoothed periodogram (——) of the female residuals. (c) The unsmoothed (+) and smoothed (——) estimates of the co-spectrum. (d) The unsmoothed (+) and smoothed (——) estimates of the quadrature spectrum.

8.6 Further reading

More detailed accounts of bivariate and, more generally, multivariate time-series analysis are available in a number of books including Hannan (1970) and the second volume of Priestley (1981). Both Hannan and Priestley emphasize frequency domain methods. Box and Jenkins (1970,

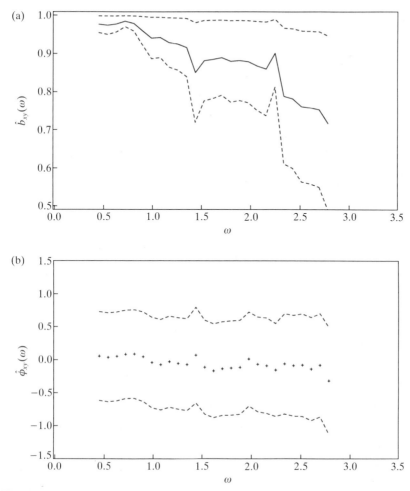

Fig. 8.12. The estimates of coherency and phase derived from the smoothed estimates shown in Fig. 8.11. (a) estimated coherency (——) and approximate pointwise 95% confidence limits (– – –). (b) Estimated phase (+) and approximate 95% confidence limits (– – –).

Chapters 10, 11) discuss time-domain methods for transfer function modelling, in which an output series y_t is related to one or more input series x_t by a system of linear equations. Systems of linear equations feature prominently in econometric modelling (Zellner 1979; Judge *et al.* 1980). Many authors have considered the use of multivariate ARIMA processes as models for multivariate time-series. See, for example, Tiao and Tsay (1989).

References

Abraham, B. and Ledolter, J. (1983). *Statistical methods for forecasting.* Wiley, New York.

Altham, P. M. E. (1984). Improving the precision of estimation by fitting a model. *Journal of the Royal Statistical Society, B* **46**, 118–9.

Anderson, O. D. (1976). *Time series analysis and forecasting: the Box–Jenkins approach.* Butterworth, London.

Bartlett, M. S. (1946). On the theoretical specification of sampling properties of autocorrelated time series. *Journal of the Royal Statistical Society, B* **8**, 27–41.

Bartlett, M. S. (1950). Periodogram analysis and continuous spectra. *Biometrika* **37**, 1–16.

Bartlett, M. S. (1954). Problèmes de l'analyse spectrale des series temporelle stationnaires. *Publications de l'Institut de Statistique de l'Université de Paris,* II (3), 119–34.

Bartlett, M. S. (1955). *Stochastic processes.* Cambridge University Press, Cambridge.

Beale, E. M. L. and Little, R. J. A. (1975). Missing values in multivariate analysis. *Journal of the Royal Statistical Society, B* **37**, 129–45.

Becker, R. A., Chambers, J. M., and Wilks, A. R. (1988). *The new S language.* Wadsworth and Brooks, Pacific Grove.

Blackman, R. B. and Tukey, J. W. (1959). *The measurement of power spectra.* Dover, New York.

Bloomfield, P. (1973). An exponential model for the spectrum of a scalar time series. *Biometrika* **60**, 217–26.

Bloomfield, P. (1976). *Fourier analysis of time series: an introduction.* Wiley, New York.

Box, G. E. P. (1950). Problems in the analysis of growth and wear curves. *Biometrics* **6**, 362–87.

Box, G. E. P. and Cox, D. R. (1964). An analysis of transformations (with Discussion). *Journal of the Royal Statistical Society, B* **26**, 211–52.

Box, G. E. P. and Jenkins, G. M. (1970). *Time series analysis, forecasting and control.* Holden-Day, San Francisco.

Box, G. E. P. and Pierce, D. A. (1970). Distribution of residual autocorrelations in autoregressive-integrated moving average time series models. *Journal of the American Statistical Association* **15**, 1509–26.

Brillinger, D. R. and Krishnaiah, P. R. (1983). *Handbook of statistics 3: time series in the frequency domain.* North-Holland, Amsterdam.

Bulmer, M. G. (1974). A statistical analysis of the ten year cycle in Canada. *Journal of Animal Ecology* **43**, 701–18.

Cameron, M. A. and Turner, T. R. (1987). Fitting models to spectra using regression packages. *Applied Statistics* **36**, 47–57.

Campbell, M. J. and Walker, A. M. (1977). A survey of statistical work on the

MacKenzie River series of annual Canadian lynx trappings for the years 1821–1934, and a new analysis. *Journal of the Royal Statistical Society, A* **140,** 411–31.

Chatfield, C. and Collins, A. J. (1980). *Introduction to multivariate analysis.* Chapman and Hall, London.

Cleveland, W. S. (1979). Robust locally weighted regression and smoothing scatterplots. *Journal of the American Statistical Association* **74,** 829–36.

Cleveland, W. S., McGill, M. E. and McGill, R. (1988). The shape parameter of a two-variable graph *Journal of the American Statistical Association* **83,** 289–300.

Coates, D. S. and Diggle, P. J. (1986). Tests for comparing two estimated spectral densities. *Journal of Time Series Analysis* **7,** 7–20.

Cooley, J. W. and Tukey, J. W. (1965). An algorithm for the machine calculation of complex Fourier series. *Mathematics of Computation* **19,** 297–301.

Copson, E. T. (1935). *An introduction to the theory of functions of a complex variable.* Oxford University Press, Oxford.

Cox, D. R. and Lewis, P. A. W. (1966). *Statistical analysis of series of events.* Chapman and Hall, London.

Cox, D. R. and Miller, H. D. (1965). *The theory of stochastic processes.* Methuen, London.

Craven, P. and Wahba, G. (1979). Smoothing noisy data with spline functions. *Numerische Mathematik* **31,** 377–403.

Crépeau, H., Koziol, J., Reid, N., and Yuh, Y. S. (1985). Analysis of incomplete multivariate data from repeated measurement experiments. *Biometrics* **41,** 505–14.

Cullis, B. R. and McGilchrist, C. A. (1989). A flexible model for the analysis of growth data from designed field experiments. *Biometrics* **45.**

Daniell, P. J. (1946). Contribution to the Discussion of Prof. Bartlett's paper. *Journal of the Royal Statistical Society supplementary* **8,** 27.

Diggle, P. J. (1988). An approach to the analysis of repeated measurements. *Biometrics* **44,** 959–71.

Diggle, P. J. and Donnelly, J. B. (1989). A selected bibliography on the analysis of repeated measurements and related areas. *Australian Journal of Statistics* **31,** 183–93.

Diggle, P. J. and Hutchinson, M. F. (1989). On spline smoothing with autocorrelated errors. *Australian Journal of Statistics* **31,** 166–82.

Diggle, P. J. and Zeger, S. L. (1989). A non-Gaussian model for time series with pulses. *Journal of the American Statistical Association* **84.**

Draper, N. R. and Smith, H. (1981). *Applied regression analysis,* 2nd edn. Wiley, New York.

Durbin, J. (1960). The fitting of time series models. *International Statistical Review* **28,** 233–44.

Elston, C. and Nicholson, M. (1942). The ten year cycle in numbers of the lynx in Canada. *Journal of Animal Ecology* **11,** 215–44.

Fisher, R. A. (1929). Tests of significance in harmonic analysis. *Proceedings of the Royal Society, A* **125,** 54–9.

Gabriel, K. R. and Neumann, J. (1962). A Markov chain model for daily rainfall occurrence at Tel Aviv. *Quarterly Journal of the Royal Meteorology Society* **88**, 90–5.

Gause, G. F. (1934). *The struggle for existence.* Williams and Williams, Baltimore.

Gilchrist, W. G. (1976). *Statistical forecasting.* Wiley, Chichester.

Glasbey, C. A. (1979). Correlated residuals in non-linear regression applied to growth data. *Applied Statistics* **28**, 251–9.

Hannan, E. J. (1970). *Multiple time series.* Wiley, New York.

Hannan, E. J. (1973). The asymptotic theory of linear time series models. *Journal of Applied Probability* **10**, 130–45.

Harrison, P. J. and Stevens, C. F. (1976). Bayesian forecasting (with Discussion). *Journal of the Royal Statistical Society, B* **38**, 205–47.

Hart, J. D. and Wehrly, T. E. (1986). Kernel regression estimation using repeated measurements data. *Journal of the American Statistical Association* **81**, 1080–8.

Hawkins, F. M. and Hawkins, J. Q. (1969). *Complex numbers and elementary complex analysis.* Gordon and Breach, New York.

Hutchinson, M. F. and De Hoog, F. R. (1985). Smoothing noisy data with spline functions. *Numerische Mathematik* **47**, 99–106.

Jameson, G. J. O. (1970). *A first course on complex functions.* Chapman and Hall, London.

Jeffreys, H. and Jeffreys, B. (1956). *Methods of mathematical physics,* 3rd edn. Cambridge University Press, Cambridge.

Jenkins, G. M. (1979). *Practical experiences with modelling and forecasting time series.* G. Jenkins, St. Helier.

Jenkins, G. M. and Watts, D. G. (1968). *Spectral analysis and its applications.* Holden-Day, San Francisco.

Jones, R. H. (1980). Maximum likelihood fitting of ARMA models to time series with missing observations. *Technometrics* **22**, 389–95.

Journel, A. G. and Huijbregts, C. J. (1978). *Mining geostatistics.* Academic Press, London.

Judge, G. C., Griffiths, W. E., Hill, R. C. and Lee, R. C. (1980). *The theory and practice of econometrics.* Wiley, New York.

Kalman, R. E. (1960). A new approach to linear filtering and prediction problems. *Journal of Basic Engineering* **82**, 34–45.

Kitagawa, G. (1987). Non-Gaussian state-space modelling of nonstationary time series. *Journal of the American Statistical Association* **82**, 1032–41.

Kleiner, B., Martin, R. D. and Thomson, D. J. (1979). Robust estimation of power spectra (with Discussion). *Journal of the Royal Statistical Society, B* **41**, 313–51.

Koch, G. G., Amara, I. A., Stokes, M. E. and Gillings, D. B. (1980). Some views on parametric and non-parametric analysis for repeated measurements and selected bibliography. *International Statistical Review* **48**, 249–65.

Koch, G. C., Landis, J. R., Freeman, J. L., Freeman, D. H. and Lehnen, R. G. (1977). A general methodology for the analysis of designed experiments with repeated measurement of categorical data. *Biometrics* **33**, 133–58.

Koopmans, L. H. (1974). *The spectral analysis of time series*. Academic Press, New York.

Lawrance, A. J. and Lewis, P. A. W. (1985). Modelling and residual analysis of nonlinear autoregressive time series in exponential variables (with Discussion). *Journal of the Royal Statistical Society, B* **47**, 165–202.

Ledermann, W. (1980). *Complex numbers*. Routledge and Kegan Paul, London.

Lincoln, D. W., Fraser, H. M., Lincoln, G. A., Martin, G. B. and McNeilly, A. S. (1985). Hypothalmic pulse generators. *Recent Progress in Hormone Research* **41**, 369–419.

Ljung, G. M. and Box, G. E. P. (1978). On a measure of lack of fit in time series models. *Biometrika* **65**, 297–303.

McCullagh, P. and Nelder, J. A. (1983). *Generalized linear models*. Chapman and Hall, London.

Makridakis, S., Andersen, A., Carbone, R., Fildes, R., Hibon, M., Lewandowski, R., Newton, J., Parzen, E., and Winkler, R. (1984). *The forecasting accuracy of major time series methods*. Wiley, Chichester.

Mardia, K. V., Kent, J. T. and Bibby, J. M. (1979). *Multivariate analysis*. Academic Press, London.

Martin, R. D. and Raftery, A. E. (1987). Comment: robustness, computation and non-Euclidean models. *Journal of the American Statistical Association* **82**, 1044–50.

Meinhold, R. J. and Singpurwalla, N. D. (1983). Understanding the Kalman filter. *American Statistician* **37**, 123–7.

Moran, P. A. P. (1953). The statistical analysis of the Canadian lynx cycle, I. *Australian Journal of Zoology* **1**, 163–73.

Murdoch, A. P., Diggle, P. J., Dunlop, W. and Kendall-Taylor, P. (1985). Determination of the frequency of pulsatile luteinizing hormone secretion by time series analysis. *Clinical Endocrinology* **22**, 341–6.

Nelder, J. A. and Mead, R. (1965). A simplex method for function minimisation. *The Computer Journal* **7**, 303–13.

Newbold, P. (1974). The exact likelihood function for a mixed autoregressive moving average process. *Biometrika* **61**, 423–6.

Nicholls, D. F. and Quinn, B. G. (1982). *Random coefficient autoregressive models: an introduction*. Springer-Verlag, New York.

Parzen, E. (1961). Mathematical considerations in the analysis of spectra. *Technometrics* **3**, 167–190.

Payne, R. W. (1987). *GENSTAT 5 reference manual*. Clarendon Press, Oxford.

Potscher, B. M. & Reschenhofer, E. (1988). Discriminating between two spectral densities in case of replicated observations. *Journal of Time Series Analysis* **9**, 221–4.

Priestley, H. A. (1985). *Introduction to complex analysis*. Oxford University Press, Oxford.

Priestley, M. B. (1981). *Spectral analysis and time series*. Academic Press, London.

Quenouille, M. H. (1949). Approximate tests of correlation in time series. *Journal of the Royal Statistical Society, B* **11**, 68–84.

Ratowsky, D. A. (1983). *Non-linear regression modelling: a unified practical approach*. Marcel Dekker, New York.

Reinsch, C. (1967). Smoothing by spline functions. *Numerische Mathematik* **10**, 177–83.

Sandland, R. L. and McGilchrist, C. A. (1979). Stochastic growth curve analysis. *Biometrics* **35**, 255–72.

Seber, G. A. F. (1977). *Regression analysis*. Wiley, New York.

Silverman, B. W. (1984a). Spline smoothing: the equivalent variable kernel method. *Annals of Statistics* **12**, 898–918.

Silverman, B. W. (1984b). A fast and efficient cross-validation method for smoothing parameter choice in spline regression. *Journal of the American Statistical Association* **79**, 584–9.

Silverman, B. W. (1986). *Density estimation for statistics and data analysis*. Chapman and Hall, London.

Smith, A. F. M. and West, M. (1983). Monitoring renal transplants: an application of the multiprocess Kalman filter. *Biometrics* **39**, 867–78.

SPSS (1988). *SPSS-X User's Guide,* 3rd edn. SPSS Inc, Chicago.

Stephens, M. A. (1974). EDF statistics for goodness of fit and some comparisons. *Journal of the American Statistical Association* **69**, 730–7.

Stern, R. D. and Coe, R. (1984). A model fitting analysis of daily rainfall data (with Discussion). *Journal of the Royal Statistical Society, A* **147**, 1–34.

Stiratelli, R., Laird, N. and Ware, J. H. (1984). Random-effects models for serial observations with binary response. *Biometrics* **40**, 961–71.

Thompson, W. A. (1962). The problem of negative estimates of variance components. *The Annals of Mathematical Statistics* **33**, 273–89.

Tiao, G. C. and Tsay, R. S. (1989). Model specification in multivariate time series (with Discussion). *Journal of the Royal Statistical Society, B* **51**, 157–213.

Tong, H. (1977). Some comments on the Canadian Lynx data. *Journal of the Royal Statistical Society, A* **140**, 432–6.

Tufte, E. R. (1983). *The visual display of quantitative information*. Graphics Press, Cheshire.

Tukey, J. W. (1949). The sampling theory of power spectrum estimates. *Proceedings on Applications of Autocovariance Analysis to Physical Problems,* 47–67. Office of Naval Research, Washington.

Tukey, J. W. (1977). *Exploratory data analysis*. Addison-Wesley, Reading.

Tunnicliffe-Wilson, G. (1989). On the use of marginal likelihood in time series model estimation. *Journal of the Royal Statistical Society, B* **51**, 15–27.

Velleman, P. F. (1980). Definition and comparison of robust nonlinear data smoothing algorithms. *Journal of the American Statistical Association* **75**, 609–15.

Wahba, G. (1980). Automatic smoothing of the log periodogram. *Journal of the American Statistical Association* **75**, 122–32.

Wahba, G. and Wold, S. (1975). Periodic splines for spectral density estimation; the use of cross-validation for determining the degree of smoothing. *Communications in Statistics* **4**, 125–41.

West, M., Harrison, P. J., and Migon, H. S. (1985). Dynamic generalized linear models and Bayesian forecasting. *Journal of the American Statistical Association* **80**, 73–97.

Whittle, P. (1984). *Prediction and regulation*, 2nd edn. Blackwell, Oxford.

Zeger, S. L. and Liang, K. Y. (1986). Longitudinal data analysis for discrete and continuous outcomes. *Biometrics* **42**, 121–30.

Zellner, A. (1979). Statistical analysis of econometric models. *Journal of the American Statistical Association* **74**, 628–43.

Appendix A
Data sets

Table A.1. Levels of luteinizing hormone in blood samples (Dr A. Murdoch).
Columns as follows:

1. Serial number.
2. Cycle 1, late follicular phase.
3. Cycle 2, early follicular phase.
4. Cycle 2, late follicular phase.

1	2	3	4	1	2	3	4
1	5.5	2.4	4.3	25	4.8	2.3	4.5
2	4.5	2.4	4.6	26	5.5	2.0	4.6
3	5.1	2.4	4.7	27	5.1	2.0	5.8
4	5.5	2.2	4.1	28	5.2	2.9	5.0
5	5.7	2.1	4.1	29	5.0	2.9	5.1
6	5.1	1.5	5.2	30	4.0	2.7	4.5
7	4.3	2.3	5.0	31	3.7	2.7	4.2
8	4.8	2.3	4.4	32	4.8	2.3	6.0
9	5.6	2.5	4.2	33	5.9	2.6	5.6
10	5.9	2.0	5.1	34	5.5	2.4	5.4
11	6.0	1.9	5.1	35	4.9	1.8	5.0
12	5.1	1.7	4.7	36	4.4	1.7	4.4
13	5.2	2.2	4.4	37	4.7	1.5	4.6
14	4.4	1.8	3.9	38	4.2	1.4	5.7
15	5.5	3.2	5.4	39	5.5	2.1	5.2
16	5.4	3.2	5.9	40	4.9	3.3	5.0
17	4.1	2.7	4.2	41	4.8	3.5	4.4
18	4.4	2.2	4.1	42	4.5	3.5	5.7
19	4.7	2.2	4.1	43	4.9	3.1	5.7
20	4.6	1.9	3.6	44	4.9	2.6	4.8
21	6.0	1.9	3.1	45	4.5	2.1	3.4
22	5.6	1.8	4.8	46	4.2	3.4	5.5
23	5.1	2.7	5.1	47	4.9	3.0	5.5
24	4.7	3.0	5.1	48	5.9	2.9	5.6

Table A.2. Wool prices at weekly markets (Dr J. Ive)
Columns as follows:
  1. Calendar year (1976–84).
  2. Calendar week (1–52).
  3. Weeks since 1.1.76.
  4. Floor price (cents per kilogram clean).
  5. 19mu price (cents per kilogram clean).
  6. Ratio of 19mu to floor price.
  7. Log(floor price).
  8. Log(19mu price).
  9. Log(19mu price/floor price).

1	2	3	4	5	6	7	8	9
76	33	33	262	307	1.1718	5.5683	5.7268	0.1585
76	34	34	259	306	1.1815	5.5568	5.7236	0.1668
76	35	35	257	306	1.1907	5.5491	5.7236	0.1745
76	36	36	262	306	1.1679	5.5683	5.7236	0.1553
76	37	37	264	306	1.1591	5.5759	5.7236	0.1477
76	38	38	268	313	1.1679	5.5910	5.7462	0.1552
76	39	39	276	319	1.1558	5.6204	5.7652	0.1448
76	40	40	283	324	1.1449	5.6454	5.7807	0.1353
76	41	41	290	332	1.1448	5.6699	5.8051	0.1352
76	42	42	286	320	1.1189	5.6560	5.7683	0.1123
76	43	43	288	320	1.1111	5.6630	5.7683	0.1053
76	44	44	286	314	1.0979	5.6560	5.7494	0.0934
76	45	45	279	306	1.0968	5.6312	5.7236	0.0924
76	46	46	279	307	1.1004	5.6312	5.7268	0.0956
76	47	47	280	308	1.1000	5.6348	5.7301	0.0953
76	49	49	328	358	1.0915	5.7930	5.8805	0.0875
76	50	50	334	360	1.0778	5.8111	5.8861	0.0750
77	2	54	332	360	1.0843	5.8051	5.8861	0.0810
77	3	55	332	362	1.0904	5.8051	5.8916	0.0865
77	4	56	326	361	1.1074	5.7869	5.8889	0.1020
77	5	57	322	359	1.1149	5.7746	5.8833	0.1087
77	6	58	324	359	1.1080	5.7807	5.8833	0.1026
77	7	59	322	359	1.1149	5.7746	5.8833	0.1087
77	8	60	316	357	1.1297	5.7557	5.8777	0.1220
77	9	61	314	357	1.1369	5.7494	5.8777	0.1283
77	10	62	316	357	1.1297	5.7557	5.8777	0.1220
77	11	63	316	357	1.1297	5.7557	5.8777	0.1220
77	12	64	315	357	1.1333	5.7526	5.8777	0.1251
77	13	65	314	356	1.1338	5.7494	5.8749	0.1255
77	16	68	313	354	1.1310	5.7462	5.8693	0.1231

Table A.2. (*Continued*)
1. Calendar year (1976–84).
2. Calendar week (1–52).
3. Weeks since 1.1.76.
4. Floor price (cents per kilogram clean).
5. 19mu price (cents per kilogram clean).
6. Ratio of 19mu to floor price.
7. Log(floor price).
8. Log(19mu price).
9. Log(19mu price/floor price).

1	2	3	4	5	6	7	8	9
77	17	69	312	354	1.1346	5.7430	5.8693	0.1263
77	18	70	307	354	1.1531	5.7268	5.8693	0.1425
77	19	71	305	354	1.1607	5.7203	5.8693	0.1490
77	20	72	305	354	1.1607	5.7203	5.8693	0.1490
77	21	73	299	354	1.1839	5.7004	5.8693	0.1689
77	22	74	297	354	1.1919	5.6937	5.8693	0.1756
77	23	75	300	354	1.1800	5.7038	5.8693	0.1655
77	24	76	300	354	1.1800	5.7038	5.8693	0.1655
77	25	77	302	354	1.1722	5.7104	5.8693	0.1589
77	26	78	300	354	1.1800	5.7038	5.8693	0.1655
77	33	85	293	349	1.1911	5.6802	5.8551	0.1749
77	34	86	293	349	1.1911	5.6802	5.8551	0.1749
77	35	87	293	347	1.1843	5.6802	5.8493	0.1691
77	36	88	296	347	1.1723	5.6904	5.8493	0.1589
77	37	89	295	347	1.1763	5.6870	5.8493	0.1623
77	38	90	297	347	1.1684	5.6937	5.8493	0.1556
77	39	91	298	347	1.1644	5.6971	5.8493	0.1522
77	40	92	299	346	1.1572	5.7004	5.8464	0.1460
77	41	93	301	347	1.1528	5.7071	5.8493	0.1422
77	42	94	305	348	1.1410	5.7203	5.8522	0.1319
77	43	95	307	348	1.1336	5.7268	5.8522	0.1254
77	44	96	306	346	1.1307	5.7236	5.8464	0.1228
77	45	97	303	345	1.1386	5.7137	5.8435	0.1298
77	46	98	304	345	1.1349	5.7170	5.8435	0.1265
77	47	99	303	345	1.1386	5.7137	5.8435	0.1298
77	48	100	300	345	1.1500	5.7038	5.8435	0.1397
77	49	101	299	345	1.1538	5.7004	5.8435	0.1431
77	50	102	297	345	1.1616	5.6937	5.8435	0.1498
78	2	106	297	345	1.1616	5.6937	5.8435	0.1498
78	3	107	297	345	1.1616	5.6937	5.8435	0.1498
78	4	108	298	345	1.1577	5.6971	5.8435	0.1464

Table A.2. (*Continued*)

1	2	3	4	5	6	7	8	9
78	5	109	299	345	1.1538	5.7004	5.8435	0.1431
78	6	110	303	347	1.1452	5.7137	5.8493	0.1356
78	7	111	304	348	1.1447	5.7170	5.8522	0.1352
78	8	112	304	349	1.1480	5.7170	5.8551	0.1381
78	9	113	306	350	1.1438	5.7236	5.8579	0.1343
78	10	114	307	348	1.1336	5.7268	5.8522	0.1254
78	11	115	304	347	1.1414	5.7170	5.8493	0.1323
78	15	119	305	350	1.1475	5.7203	5.8579	0.1376
78	16	120	307	353	1.1498	5.7268	5.8665	0.1397
78	17	121	307	354	1.1531	5.7268	5.8693	0.1425
78	18	122	307	355	1.1564	5.7268	5.8721	0.1453
78	19	123	308	355	1.1526	5.7301	5.8721	0.1420
78	20	124	308	355	1.1526	5.7301	5.8721	0.1420
78	21	125	309	355	1.1489	5.7333	5.8721	0.1388
78	22	126	312	355	1.1378	5.7430	5.8721	0.1291
78	23	127	315	358	1.1365	5.7526	5.8805	0.1279
78	24	128	314	358	1.1401	5.7494	5.8805	0.1311
78	25	129	310	355	1.1452	5.7366	5.8721	0.1355
78	33	137	316	356	1.1266	5.7557	5.8749	0.1192
78	34	138	314	356	1.1338	5.7494	5.8749	0.1255
78	35	139	316	356	1.1266	5.7557	5.8749	0.1192
78	36	140	314	356	1.1338	5.7494	5.8749	0.1255
78	37	141	314	356	1.1338	5.7494	5.8749	0.1255
78	38	142	316	357	1.1297	5.7557	5.8777	0.1220
78	39	143	316	357	1.1297	5.7557	5.8777	0.1220
78	40	144	316	356	1.1266	5.7557	5.8749	0.1192
78	41	145	317	358	1.1293	5.7589	5.8805	0.1216
78	42	146	317	358	1.1293	5.7589	5.8805	0.1216
78	43	147	316	358	1.1329	5.7557	5.8805	0.1248
78	44	148	315	359	1.1397	5.7526	5.8833	0.1307
78	45	149	315	359	1.1397	5.7526	5.8833	0.1307
78	46	150	315	358	1.1365	5.7526	5.8805	0.1279
78	47	151	315	357	1.1333	5.7526	5.8777	0.1251
78	48	152	316	357	1.1297	5.7557	5.8777	0.1220
78	49	153	318	360	1.1321	5.7621	5.8861	0.1240
78	50	154	318	360	1.1321	5.7621	5.8861	0.1240
79	2	158	318	358	1.1258	5.7621	5.8805	0.1184
79	3	159	319	359	1.1254	5.7652	5.8833	0.1181
79	4	160	319	363	1.1379	5.7652	5.8944	0.1292
79	5	161	320	363	1.1344	5.7683	5.8944	0.1261

Table A.2. (*Continued*)

1. Calendar year (1976–84).
2. Calendar week (1–52).
3. Weeks since 1.1.76.
4. Floor price (cents per kilogram clean).
5. 19mu price (cents per kilogram clean).
6. Ratio of 19mu to floor price.
7. Log(floor price).
8. Log(19mu price).
9. Log(19mu price/floor price).

1	2	3	4	5	6	7	8	9
79	6	162	322	371	1.1522	5.7746	5.9162	0.1416
79	7	163	330	379	1.1485	5.7991	5.9375	0.1384
79	8	164	343	400	1.1662	5.8377	5.9915	0.1538
79	9	165	354	405	1.1441	5.8693	6.0039	0.1346
79	10	166	356	412	1.1573	5.8749	6.0210	0.1461
79	11	167	366	427	1.1667	5.9026	6.0568	0.1542
79	12	168	364	427	1.1731	5.8972	6.0568	0.1596
79	13	169	371	434	1.1698	5.9162	6.0730	0.1568
79	14	170	369	448	1.2141	5.9108	6.1048	0.1940
79	18	174	364	448	1.2308	5.8972	6.1048	0.2076
79	19	175	363	469	1.2920	5.8944	6.1506	0.2562
79	20	176	362	469	1.2956	5.8916	6.1506	0.2590
79	21	177	364	469	1.2885	5.8972	6.1506	0.2534
79	22	178	361	469	1.2992	5.8889	6.1506	0.2617
79	23	179	360	469	1.3028	5.8861	6.1506	0.2645
79	24	180	362	462	1.2762	5.8916	6.1356	0.2440
79	25	181	363	462	1.2727	5.8944	6.1356	0.2412
79	26	182	361	462	1.2798	5.8889	6.1356	0.2467
79	31	187	364	462	1.2692	5.8972	6.1356	0.2384
79	34	190	369	464	1.2575	5.9108	6.1399	0.2291
79	35	191	365	464	1.2712	5.8999	6.1399	0.2400
79	36	192	366	464	1.2678	5.9026	6.1399	0.2373
79	37	193	368	469	1.2745	5.9081	6.1506	0.2425
79	38	194	370	469	1.2676	5.9135	6.1506	0.2371
79	39	195	382	490	1.2827	5.9454	6.1944	0.2490
79	40	196	408	521	1.2770	6.0113	6.2558	0.2445
79	41	197	414	536	1.2947	6.0259	6.2841	0.2582
79	42	198	400	536	1.3400	5.9915	6.2841	0.2926
79	43	199	405	533	1.3160	6.0039	6.2785	0.2746
79	44	200	397	525	1.3224	5.9839	6.2634	0.2795
79	45	201	394	519	1.3173	5.9764	6.2519	0.2755

Table A.2. (*Continued*)

1	2	3	4	5	6	7	8	9
79	46	202	397	523	1.3174	5.9839	6.2596	0.2757
79	47	203	388	517	1.3325	5.9610	6.2480	0.2870
79	48	204	379	509	1.3430	5.9375	6.2324	0.2949
79	49	205	377	507	1.3448	5.9322	6.2285	0.2963
79	50	206	380	500	1.3158	5.9402	6.2146	0.2744
80	2	210	383	510	1.3316	5.9480	6.2344	0.2864
80	3	211	394	524	1.3299	5.9764	6.2615	0.2851
80	4	212	408	538	1.3186	6.0113	6.2879	0.2766
80	5	213	413	538	1.3027	6.0234	6.2879	0.2645
80	7	215	433	573	1.3233	6.0707	6.3509	0.2802
80	13	221	415	570	1.3735	6.0283	6.3456	0.3173
80	14	222	414	566	1.3671	6.0259	6.3386	0.3127
80	15	223	414	567	1.3696	6.0259	6.3404	0.3145
80	16	224	412	584	1.4175	6.0210	6.3699	0.3489
80	17	225	405	585	1.4444	6.0039	6.3716	0.3677
80	18	226	396	564	1.4242	5.9814	6.3351	0.3537
80	19	227	396	565	1.4268	5.9814	6.3368	0.3554
80	20	228	405	573	1.4148	6.0039	6.3509	0.3470
80	22	230	407	587	1.4423	6.0088	6.3750	0.3662
80	23	231	409	585	1.4303	6.0137	6.3716	0.3579
80	24	232	403	578	1.4342	5.9989	6.3596	0.3607
80	25	233	402	578	1.4378	5.9965	6.3596	0.3631
80	26	234	401	578	1.4414	5.9940	6.3596	0.3656
80	31	239	396	576	1.4545	5.9814	6.3561	0.3747
80	35	243	398	563	1.4146	5.9865	6.3333	0.3468
80	36	244	394	557	1.4137	5.9764	6.3226	0.3462
80	37	245	396	550	1.3889	5.9814	6.3099	0.3285
80	38	246	404	564	1.3960	6.0014	6.3351	0.3337
80	39	247	408	564	1.3824	6.0113	6.3351	0.3238
80	40	248	401	564	1.4065	5.9940	6.3351	0.3411
80	41	249	398	548	1.3769	5.9865	6.3063	0.3198
80	42	250	392	535	1.3648	5.9713	6.2823	0.3110
80	43	251	393	542	1.3791	5.9738	6.2953	0.3215
80	44	252	397	536	1.3501	5.9839	6.2841	0.3002
80	45	253	400	536	1.3400	5.9915	6.2841	0.2926
80	46	254	403	540	1.3400	5.9989	6.2916	0.2927
80	47	255	416	543	1.3053	6.0307	6.2971	0.2664
80	48	256	415	544	1.3108	6.0283	6.2989	0.2706
80	49	257	414	536	1.2947	6.0259	6.2841	0.2582
80	50	258	414	526	1.2705	6.0259	6.2653	0.2394

Table A.2. (*Continued*)

1. Calendar year (1976–84).
2. Calendar week (1–52).
3. Weeks since 1.1.76.
4. Floor price (cents per kilogram clean).
5. 19mu price (cents per kilogram clean).
6. Ratio of 19mu to floor price.
7. Log(floor price).
8. Log(19mu price).
9. Log(19mu price/floor price).

1	2	3	4	5	6	7	8	9
81	3	263	415	528	1.2723	6.0283	6.2691	0.2408
81	4	264	425	530	1.2471	6.0521	6.2729	0.2208
81	5	265	427	534	1.2506	6.0568	6.2804	0.2236
81	6	266	426	526	1.2347	6.0544	6.2653	0.2109
81	7	267	421	525	1.2470	6.0426	6.2634	0.2208
81	8	268	420	525	1.2500	6.0403	6.2634	0.2231
81	9	269	419	524	1.2506	6.0379	6.2615	0.2236
81	10	270	417	523	1.2542	6.0331	6.2596	0.2265
81	11	271	411	519	1.2628	6.0186	6.2519	0.2333
81	12	272	406	519	1.2783	6.0064	6.2519	0.2455
81	13	273	410	533	1.3000	6.0162	6.2785	0.2623
81	14	274	409	540	1.3203	6.0137	6.2916	0.2779
81	15	275	411	540	1.3139	6.0186	6.2916	0.2730
81	18	278	413	543	1.3148	6.0234	6.2971	0.2737
81	19	279	415	543	1.3084	6.0283	6.2971	0.2688
81	20	280	419	546	1.3031	6.0379	6.3026	0.2647
81	21	281	423	541	1.2790	6.0474	6.2934	0.2460
81	22	282	422	539	1.2773	6.0450	6.2897	0.2447
81	23	283	425	539	1.2682	6.0521	6.2897	0.2376
81	25	285	428	539	1.2593	6.0591	6.2897	0.2306
81	26	286	427	543	1.2717	6.0568	6.2971	0.2403
81	31	291	427	543	1.2717	6.0568	6.2971	0.2403
81	35	295	425	543	1.2776	6.0521	6.2971	0.2450
81	36	296	423	518	1.2246	6.0474	6.2500	0.2026
81	37	297	422	519	1.2299	6.0450	6.2519	0.2069
81	39	299	422	519	1.2299	6.0450	6.2519	0.2069
81	40	300	420	519	1.2357	6.0403	6.2519	0.2116
81	41	301	421	522	1.2399	6.0426	6.2577	0.2151
81	42	302	421	523	1.2423	6.0426	6.2596	0.2170
81	43	303	418	520	1.2440	6.0355	6.2538	0.2183
81	44	304	421	522	1.2399	6.0426	6.2577	0.2151

Table A.2. (*Continued*)

1	2	3	4	5	6	7	8	9
81	45	305	420	520	1.2381	6.0403	6.2538	0.2135
81	46	306	422	524	1.2417	6.0450	6.2615	0.2165
81	47	307	424	523	1.2335	6.0497	6.2596	0.2099
81	48	308	425	523	1.2306	6.0521	6.2596	0.2075
81	49	309	423	522	1.2340	6.0474	6.2577	0.2103
81	50	310	422	522	1.2370	6.0450	6.2577	0.2127
81	51	311	422	524	1.2417	6.0450	6.2615	0.2165
82	3	315	423	524	1.2388	6.0474	6.2615	0.2141
82	4	316	423	530	1.2530	6.0474	6.2729	0.2255
82	5	317	427	535	1.2529	6.0568	6.2823	0.2255
82	6	318	428	535	1.2500	6.0591	6.2823	0.2232
82	7	319	437	544	1.2449	6.0799	6.2989	0.2190
82	8	320	441	550	1.2472	6.0890	6.3099	0.2209
82	9	321	440	550	1.2500	6.0868	6.3099	0.2231
82	10	322	445	554	1.2449	6.0981	6.3172	0.2191
82	11	323	446	561	1.2578	6.1003	6.3297	0.2294
82	12	324	443	579	1.3070	6.0936	6.3613	0.2677
82	13	325	445	579	1.3011	6.0981	6.3613	0.2632
82	14	326	444	582	1.3108	6.0958	6.3665	0.2707
82	18	330	446	585	1.3117	6.1003	6.3716	0.2713
82	19	331	445	581	1.3056	6.0981	6.3648	0.2667
82	20	332	443	581	1.3115	6.0936	6.3648	0.2712
82	21	333	442	581	1.3145	6.0913	6.3648	0.2735
82	22	334	446	580	1.3004	6.1003	6.3630	0.2627
82	23	335	448	579	1.2924	6.1048	6.3613	0.2565
82	24	336	449	579	1.2895	6.1070	6.3613	0.2543
82	25	337	445	579	1.3011	6.0981	6.3613	0.2632
82	26	338	443	579	1.3070	6.0936	6.3613	0.2677
82	27	339	437	579	1.3249	6.0799	6.3613	0.2814
82	30	342	439	579	1.3189	6.0845	6.3613	0.2768
82	34	346	433	572	1.3210	6.0707	6.3491	0.2784
82	35	347	431	572	1.3271	6.0661	6.3491	0.2830
82	36	348	430	572	1.3302	6.0638	6.3491	0.2853
82	37	349	437	576	1.3181	6.0799	6.3561	0.2762
82	38	350	435	576	1.3241	6.0753	6.3561	0.2808
82	39	351	433	561	1.2956	6.0707	6.3297	0.2590
82	40	352	432	561	1.2986	6.0684	6.3297	0.2613
82	41	353	432	553	1.2801	6.0684	6.3154	0.2470
82	42	354	431	558	1.2947	6.0661	6.3244	0.2583
82	43	355	432	553	1.2801	6.0684	6.3154	0.2470

Table A.2. (*Continued*)
1. Calendar year (1976–84).
2. Calendar week (1–52).
3. Weeks since 1.1.76.
4. Floor price (cents per kilogram clean).
5. 19mu price (cents per kilogram clean).
6. Ratio of 19mu to floor price.
7. Log(floor price).
8. Log(19mu price).
9. Log(19mu price/floor price).

1	2	3	4	5	6	7	8	9
82	44	356	432	547	1.2662	6.0684	6.3044	0.2360
82	45	357	432	562	1.3009	6.0684	6.3315	0.2631
82	46	358	432	565	1.3079	6.0684	6.3368	0.2684
82	47	359	430	564	1.3116	6.0638	6.3351	0.2713
82	48	360	427	551	1.2904	6.0568	6.3117	0.2549
82	49	361	426	549	1.2887	6.0544	6.3081	0.2537
82	50	362	426	549	1.2887	6.0544	6.3081	0.2537
83	2	366	429	556	1.2960	6.0615	6.3208	0.2593
83	3	367	432	563	1.3032	6.0684	6.3333	0.2649
83	4	368	435	579	1.3310	6.0753	6.3613	0.2860
83	5	369	438	580	1.3242	6.0822	6.3630	0.2808
83	6	370	438	582	1.3288	6.0822	6.3665	0.2843
83	7	371	436	577	1.3234	6.0776	6.3578	0.2802
83	8	372	435	578	1.3287	6.0753	6.3596	0.2843
83	9	373	435	576	1.3241	6.0753	6.3561	0.2808
83	10	374	471	626	1.3291	6.1549	6.4394	0.2845
83	11	375	471	628	1.3333	6.1549	6.4425	0.2876
83	12	376	466	641	1.3755	6.1442	6.4630	0.3188
83	13	377	467	623	1.3340	6.1463	6.4345	0.2882
83	17	381	470	639	1.3596	6.1527	6.4599	0.3072
83	18	382	470	643	1.3681	6.1527	6.4661	0.3134
83	19	383	470	632	1.3447	6.1527	6.4489	0.2962
83	20	384	468	632	1.3504	6.1485	6.4489	0.3004
83	21	385	467	632	1.3533	6.1463	6.4489	0.3026
83	22	386	472	635	1.3453	6.1570	6.4536	0.2966
83	23	387	473	635	1.3425	6.1591	6.4536	0.2945
83	24	388	474	644	1.3586	6.1612	6.4677	0.3065
83	25	389	471	642	1.3631	6.1549	6.4646	0.3097
83	30	394	481	638	1.3264	6.1759	6.4583	0.2824
83	34	398	481	627	1.3035	6.1759	6.4409	0.2650
83	35	399	481	627	1.3035	6.1759	6.4409	0.2650

Table A.2. (*Continued*)

1	2	3	4	5	6	7	8	9
83	36	400	481	624	1.2973	6.1759	6.4362	0.2603
83	37	401	482	627	1.3008	6.1779	6.4409	0.2630
83	38	402	482	627	1.3008	6.1779	6.4409	0.2630
83	39	403	482	633	1.3133	6.1779	6.4505	0.2726
83	40	404	481	624	1.2973	6.1759	6.4362	0.2603
83	41	405	480	620	1.2917	6.1738	6.4297	0.2559
83	42	406	479	617	1.2881	6.1717	6.4249	0.2532
83	43	407	477	620	1.2998	6.1675	6.4297	0.2622
83	44	408	477	625	1.3103	6.1675	6.4378	0.2703
83	45	409	477	620	1.2998	6.1675	6.4297	0.2622
83	46	410	476	624	1.3109	6.1654	6.4362	0.2708
83	47	411	476	623	1.3088	6.1654	6.4345	0.2691
83	48	412	476	627	1.3172	6.1654	6.4409	0.2755
83	49	413	475	628	1.3221	6.1633	6.4425	0.2792
83	50	414	476	636	1.3361	6.1654	6.4552	0.2898
84	2	418	480	649	1.3521	6.1738	6.4754	0.3016
84	3	419	483	671	1.3892	6.1800	6.5088	0.3288
84	4	420	488	717	1.4693	6.1903	6.5751	0.3848
84	5	421	490	748	1.5265	6.1944	6.6174	0.4230
84	6	422	488	726	1.4877	6.1903	6.5876	0.3973
84	7	423	490	754	1.5388	6.1944	6.6254	0.4310
84	8	424	490	754	1.5388	6.1944	6.6254	0.4310
84	9	425	493	791	1.6045	6.2005	6.6733	0.4728
84	10	426	490	775	1.5816	6.1944	6.6529	0.4585
84	11	427	490	773	1.5776	6.1944	6.6503	0.4559
84	12	428	487	770	1.5811	6.1883	6.6464	0.4581
84	13	429	492	786	1.5976	6.1985	6.6670	0.4685
84	14	430	495	786	1.5879	6.2046	6.6670	0.4624
84	18	434	494	786	1.5911	6.2025	6.6670	0.4645
84	19	435	496	786	1.5847	6.2066	6.6670	0.4604
84	20	436	495	786	1.5879	6.2046	6.6670	0.4624
84	21	437	495	786	1.5879	6.2046	6.6670	0.4624
84	22	438	497	786	1.5815	6.2086	6.6670	0.4584
84	25	441	496	786	1.5847	6.2066	6.6670	0.4604
84	26	442	499	786	1.5752	6.2126	6.6670	0.4544

Table A.3. U.K. deaths from bronchitis, emphysema, and asthma (Dr. D. Appleton). The data are the numbers of deaths reported monthly, for the years 1974 to 1979 inclusive. Numbers are recorded separately for males and for females.

(a) males (read row-wise)

2134	1863	1877	1877	1492	1249	1280	1131	1209	1492
1621	1846	2103	2137	2153	1833	1403	1288	1186	1133
1053	1347	1545	2066	2020	2750	2283	1479	1189	1160
1113	970	999	1208	1467	2059	2240	1634	1722	1801
1246	1162	1087	1013	959	1179	1229	1655	2019	2284
1942	1423	1340	1187	1098	1004	970	1140	1110	1812
2263	1820	1846	1531	1215	1075	1056	975	940	1081
1294	1341								

(b) females (read row-wise)

901	689	827	677	522	406	441	393	387	582
578	666	830	752	785	664	467	438	421	412
343	440	531	771	767	1141	896	532	447	420
376	330	357	445	546	764	862	660	663	643
502	392	411	348	387	385	411	638	796	853
737	546	530	446	431	362	387	430	425	679
821	785	727	612	478	429	405	379	393	411
487	574								

Table A.4. Growth of colonies of *paramecium aurelium* (Gause 1934). Columns as follows:

1. Number of days since start of experiment.
2,3,4. Numbers of sampled organisms in three replicates.

1	2	3	4
0	2	2	2
2	17	15	11
3	29	36	37
4	39	62	67
5	63	84	134
6	185	156	226
7	258	234	306
8	267	348	376
9	392	370	485
10	510	480	530
11	570	520	650
12	650	575	605
13	560	400	580
14	575	545	660
15	650	560	460
16	550	480	650
17	480	510	575
18	520	650	525
19	500	500	550

Table A.5. Body-weights of rats under three different experimental experimental treatments (Box 1950). Body-weights are recorded at five, weekly intervals for 27 rats in three experimental treatment groups.

(a) Controls

57	86	114	139	172
60	93	123	146	177
52	77	111	144	185
49	67	100	129	164
56	81	104	121	151
46	70	102	131	153
51	71	94	110	141
63	91	112	130	154
49	67	90	112	140
57	82	110	139	169

(b) Thyroxin

59	85	121	146	181
54	71	90	110	138
56	75	108	151	189
59	85	116	148	177
57	72	97	120	144
52	73	97	116	140
52	70	105	138	171

(c) Thiourocil

61	86	109	120	129
59	80	101	111	122
53	79	100	106	133
59	88	100	111	122
51	75	101	123	140
51	75	92	100	119
56	78	95	103	108
58	69	93	116	140
46	61	78	90	107
53	72	89	104	122

Table A.6. Protein content of milk samples (Ms. A. Frensham). Percentage protein content of milk samples recorded at 19, weekly intervals from 79 cows on three different diets. Missing values are indicated by zeros

(a) Barley (25 cows)

3.63	3.57	3.47	3.65	3.89	3.73	3.77	3.90	3.78	3.82	3.83	3.71	4.10	4.02	4.13	4.08	4.22	4.44	4.30
3.24	3.25	3.29	3.09	3.38	3.33	3.00	3.16	3.34	3.32	3.31	3.27	3.41	3.45	3.12	3.42	3.40	3.17	3.00
3.98	3.60	3.43	3.30	3.29	3.25	2.93	3.20	3.27	3.22	2.93	2.92	2.82	2.64	0.00	0.00	0.00	0.00	0.00
3.66	3.50	3.05	2.90	2.72	3.11	3.05	2.80	3.20	3.18	3.14	3.18	3.24	3.37	3.30	3.40	3.35	3.28	0.00
4.34	3.76	3.68	3.51	3.45	3.53	3.60	3.77	3.90	3.87	3.61	3.85	3.94	3.87	3.60	3.06	3.47	3.50	3.42
4.36	3.71	3.42	3.95	4.06	3.73	3.92	3.99	3.70	3.88	3.71	3.62	3.74	3.42	0.00	0.00	0.00	0.00	0.00
4.17	3.60	3.52	3.10	3.78	3.42	3.66	3.64	3.83	3.73	3.72	3.65	3.50	3.32	2.95	3.34	3.51	3.17	0.00
4.40	3.86	3.56	3.32	3.64	3.57	3.47	3.97	0.00	3.78	3.98	3.90	4.05	4.06	4.05	3.92	3.65	3.60	3.74
3.40	3.42	3.51	3.39	3.35	3.13	3.21	3.50	3.55	3.28	3.75	3.55	3.53	3.52	3.77	3.77	3.74	4.00	3.87
3.75	3.89	3.65	3.42	3.32	3.27	3.34	3.35	3.09	3.65	3.53	3.50	3.63	3.91	3.73	3.71	4.18	3.97	4.06
4.20	3.59	3.55	3.27	3.19	3.60	3.50	3.55	3.60	3.75	3.75	3.75	3.89	3.87	3.60	3.68	3.68	3.56	3.34
4.02	3.76	3.60	3.53	3.95	3.26	3.73	3.96	0.00	3.70	0.00	3.45	3.50	3.13	0.00	0.00	0.00	0.00	0.00
4.02	3.90	3.73	3.55	3.71	3.40	3.49	3.74	3.61	3.42	3.46	3.40	3.38	3.13	0.00	0.00	0.00	0.00	0.00
3.90	3.33	3.25	3.22	3.35	3.24	3.16	3.33	3.12	2.93	2.84	3.07	3.02	2.75	0.00	0.00	0.00	0.00	0.00
3.81	4.00	3.57	3.47	3.52	3.63	3.45	3.50	3.71	3.55	3.13	3.04	3.31	3.22	2.92	3.02	0.00	0.00	0.00
3.62	3.22	3.62	3.02	3.28	3.15	3.52	3.22	3.45	3.51	3.38	3.00	2.94	3.52	3.48	3.48	0.00	0.00	0.00
3.66	3.66	3.28	3.10	2.66	3.00	3.15	3.01	3.50	3.29	3.16	3.33	3.50	3.46	3.48	3.98	3.70	3.36	3.55
4.44	3.85	3.55	3.22	3.40	3.28	3.42	3.35	3.01	3.55	3.70	3.73	3.65	3.78	3.82	3.75	3.95	3.85	3.72
4.23	3.75	3.82	3.60	4.09	3.84	3.62	3.36	3.65	3.41	3.15	3.68	3.54	3.75	3.72	4.05	3.60	3.88	3.98
3.82	0.00	3.27	3.33	3.25	2.97	3.57	3.43	3.50	3.58	3.70	3.55	3.58	3.70	3.60	3.42	3.33	3.53	3.40
3.53	3.10	3.48	3.35	3.35	3.65	3.65	3.56	3.27	3.61	3.66	3.47	3.34	3.32	3.22	3.18	0.00	0.00	0.00
4.47	3.86	3.34	3.49	3.74	3.24	3.71	3.46	3.88	3.60	4.00	3.83	3.80	4.12	3.98	3.77	3.52	3.50	3.42
3.93	3.79	3.68	3.58	3.76	3.66	3.57	3.85	3.75	3.37	3.00	3.24	3.44	3.23	0.00	0.00	0.00	0.00	0.00
3.27	3.84	3.46	3.44	3.40	3.50	3.63	3.47	3.32	3.47	3.40	3.27	3.74	3.76	3.68	3.68	3.93	3.80	3.52
3.32	3.61	3.25	3.48	3.58	3.47	3.60	3.51	3.74	3.50	3.08	2.77	3.22	3.35	3.14	0.00	0.00	0.00	0.00

Table A.6. (*Continued*)

(b) Mixed diet of barley and lupins (27 cows)

3.38	3.38	3.10	3.09	3.15	2.77	3.40	3.25	3.35	3.57	3.80	3.70	3.44	3.51	3.78	3.52	3.26	3.52	3.25
3.80	3.51	3.19	3.11	3.35	3.46	3.20	3.52	3.66	3.87	3.85	3.90	4.06	3.93	3.40	3.85	3.93	3.52	3.62
4.17	3.71	3.32	3.10	3.07	3.11	3.29	3.00	3.32	3.25	3.50	3.31	3.51	3.54	3.45	3.59	3.38	3.44	3.18
4.59	3.86	3.62	3.60	3.65	3.75	3.71	3.73	3.62	3.63	3.40	3.44	3.45	3.20	0.00	0.00	0.00	0.00	0.00
4.07	3.45	3.56	3.10	3.92	3.20	3.28	3.36	3.28	3.25	2.99	3.10	3.05	2.70	0.00	0.00	0.00	0.00	0.00
4.32	3.37	3.47	3.46	3.31	3.57	3.38	3.01	3.66	3.46	3.21	3.76	3.44	3.47	3.45	3.89	3.65	3.59	3.80
3.56	3.14	3.60	3.36	3.37	3.44	3.57	3.55	3.57	3.68	3.58	3.65	3.29	3.53	3.26	2.89	0.00	0.00	0.00
3.67	3.33	3.20	2.72	2.95	3.07	2.90	3.56	3.19	3.35	3.26	3.40	3.18	3.43	3.57	3.49	3.59	3.27	3.48
4.15	3.55	3.27	3.37	3.65	3.58	3.55	3.47	3.48	3.55	3.30	3.32	3.22	2.97	0.00	0.00	0.00	0.00	0.00
3.51	3.90	2.75	3.37	3.51	3.64	3.55	3.60	3.70	3.60	3.65	3.55	3.30	3.15	2.90	0.00	0.00	0.00	0.00
4.20	3.31	3.30	3.32	3.29	3.17	3.04	3.09	3.14	3.04	3.46	3.12	3.30	3.09	3.45	3.47	3.30	3.45	3.42
4.12	3.56	3.58	3.47	3.20	3.50	3.40	3.56	3.02	3.25	3.44	3.52	3.72	3.55	3.65	3.30	3.35	3.20	2.84
3.52	3.64	3.10	3.19	2.95	2.75	3.15	3.20	3.03	3.53	3.42	3.46	3.33	3.53	3.59	4.42	3.99	3.88	3.80
4.08	3.62	3.55	3.43	3.13	3.28	3.13	3.28	3.02	3.48	3.34	3.25	3.38	3.68	3.40	3.54	3.66	3.52	3.40
4.02	3.45	3.28	3.31	3.37	3.26	3.25	3.48	3.23	3.03	2.80	3.17	3.04	3.05	0.00	0.00	0.00	0.00	0.00
3.18	3.12	2.87	3.04	2.88	3.08	3.15	3.22	3.20	3.46	3.33	3.26	3.40	3.36	3.97	3.20	3.10	3.19	3.10
4.11	4.09	3.54	3.15	3.57	3.23	3.10	3.30	3.60	3.74	3.55	3.66	3.68	3.81	3.97	3.78	3.41	3.15	3.32
3.27	3.10	3.14	2.80	3.34	3.11	3.17	3.15	3.52	3.35	3.30	3.55	0.00	3.11	2.94	3.25	3.05	2.95	0.00
3.27	3.40	3.50	3.38	3.29	3.60	3.50	3.33	3.66	3.68	3.20	3.10	3.32	3.26	3.08	0.00	0.00	0.00	0.00
3.97	3.60	3.50	3.50	3.58	3.58	3.57	3.65	3.71	3.37	3.15	3.20	3.32	3.01	0.00	0.00	0.00	0.00	0.00
3.31	3.42	3.24	3.38	3.39	3.70	3.47	3.49	3.64	3.50	3.27	2.95	3.04	3.21	3.04	3.12	3.04	3.45	0.00
4.12	3.25	3.14	3.00	3.20	3.55	3.45	3.63	3.50	3.90	3.30	3.70	3.92	3.87	3.25	0.00	0.00	0.00	3.38
3.92	3.55	3.74	3.46	3.45	3.48	3.26	3.60	3.50	2.86	3.65	3.17	3.18	3.05	0.00	3.55	3.62	3.59	0.00
3.78	3.40	3.17	3.28	3.27	3.69	3.24	3.30	3.24	2.91	3.00	2.85	3.41	3.30	3.22	0.00	0.00	0.00	3.45
4.00	3.74	3.48	3.75	3.78	3.77	3.70	3.91	3.94	3.94	3.10	3.10	3.30	3.27	0.00	4.08	0.00	0.00	0.00
4.37	4.06	3.77	3.40	3.30	3.27	3.05	3.10	3.72	3.25	3.55	3.70	3.65	3.80	3.78	3.76	4.10	3.57	0.00
3.79	4.07	3.35	3.46	3.20	3.99	3.54	0.00	3.25	3.30	3.18	3.21	3.75	3.67	3.77	3.77	3.75	3.92	3.50

(c) Lupins (27 cows)

3.53	3.25	3.12	3.03	3.46	0.00	2.66	0.00	2.78	0.00	0.00	0.00	0.00	0.00	0.00	2.71	0.00	0.00	3.40	3.65	3.52	3.18	0.00	3.50	0.00	3.09	0.00
3.77	3.42	3.30	3.34	3.63	0.00	2.75	2.95	3.17	0.00	0.00	0.00	0.00	0.00	0.00	2.90	0.00	0.00	3.70	3.29	3.20	3.31	0.00	3.45	0.00	3.35	0.00
3.78	3.28	3.09	3.38	3.56	0.00	2.89	3.15	3.13	0.00	0.00	0.00	0.00	0.00	0.00	2.96	0.00	0.00	3.26	3.19	3.04	3.32	0.00	3.60	0.00	3.15	0.00
3.78	3.41	3.18	3.10	3.68	0.00	2.76	3.23	3.21	0.00	0.00	0.00	0.00	0.00	0.00	2.75	0.00	2.96	3.30	3.58	3.42	2.92	0.00	3.49	0.00	3.47	0.00
3.70	3.40	3.20	3.60	3.64	0.00	3.08	3.17	3.10	2.68	0.00	0.00	3.17	0.00	2.87	2.84	3.20	3.15	3.51	3.40	3.27	3.18	0.00	3.40	0.00	3.71	0.00
3.74	3.17	3.31	3.60	3.45	2.93	2.95	3.43	3.05	2.88	3.00	3.22	3.35	2.88	3.05	3.14	3.35	3.29	3.26	3.36	3.13	3.75	3.20	3.21	3.04	3.87	3.25
3.77	3.30	3.22	3.63	3.37	3.24	3.04	3.55	3.07	3.03	3.18	3.31	3.30	2.79	3.04	3.20	3.52	3.00	3.17	3.58	3.20	3.75	3.37	3.51	3.20	4.00	3.67
3.59	3.26	3.26	3.60	3.18	3.28	3.15	3.64	2.87	2.70	3.17	3.31	3.14	2.97	2.67	2.87	3.29	3.18	3.04	3.52	3.38	3.59	3.33	2.90	3.14	3.55	3.19
3.42	3.25	3.00	3.49	3.42	3.22	3.13	3.21	3.15	3.23	2.94	3.29	4.22	2.55	2.57	3.02	3.57	3.60	3.08	2.88	3.25	3.45	3.22	3.15	2.97	3.55	3.45
3.60	3.38	3.05	3.28	3.00	3.45	3.25	3.30	3.15	3.32	3.07	3.51	3.76	2.45	3.11	3.20	0.00	3.76	2.80	3.05	3.00	3.55	3.46	3.12	3.25	3.78	3.34
2.97	3.20	2.87	3.59	3.44	3.33	2.80	3.47	3.17	3.44	3.38	3.74	3.90	3.00	3.35	2.90	3.96	3.60	3.02	3.05	2.88	3.60	3.84	3.00	3.35	3.60	3.92
3.22	2.85	3.13	3.27	3.37	3.40	3.12	3.37	2.85	3.41	3.43	3.90	3.75	3.12	2.95	2.86	0.00	3.50	2.77	3.01	2.94	3.55	4.13	3.10	3.50	3.61	3.95
3.07	3.12	2.87	3.11	3.02	3.38	3.08	3.30	3.08	3.52	3.25	3.57	3.38	3.00	3.03	3.06	0.00	2.83	2.74	3.13	3.05	0.00	3.65	3.27	3.20	3.27	3.70
3.30	3.05	2.84	3.00	3.36	3.28	3.00	3.49	2.92	3.54	3.30	3.33	2.72	3.26	3.15	3.24	4.30	3.45	2.82	3.10	3.14	3.64	3.90	3.35	3.34	3.30	3.45
3.09	2.82	2.92	2.79	3.45	3.57	2.61	3.41	2.84	3.65	3.45	0.00	3.52	3.30	2.83	2.74	3.48	3.31	3.20	3.06	3.46	3.10	4.20	3.10	3.40	2.94	3.95
3.50	3.07	3.17	3.25	3.61	3.50	2.84	2.88	2.99	3.65	3.45	3.43	3.38	3.31	2.92	2.86	3.45	3.45	3.22	3.11	3.06	3.37	4.00	3.14	3.20	3.28	3.85
3.00	3.37	2.80	3.34	3.82	3.62	2.78	3.39	3.02	3.84	3.55	3.52	3.42	3.10	3.05	2.96	3.43	3.53	3.33	3.58	3.20	3.47	3.93	3.44	3.38	3.35	3.85
3.38	3.35	3.04	3.34	3.61	3.70	2.89	3.51	3.30	4.12	3.40	3.59	3.75	3.00	3.35	3.12	3.52	3.19	3.12	3.42	3.39	3.37	3.85	3.81	3.22	3.26	3.95
3.69	4.20	3.31	3.13	3.73	4.32	3.04	3.84	3.98	4.18	4.20	4.10	3.25	3.34	3.50	4.13	3.21	3.90	3.50	4.10	2.69	4.30	4.06	3.88	3.95	3.67	4.27

Table A.7. Annual numbers of lynx trapped in the Mackenzie River district of North-west Canada from 1821 to 1934 inclusive (read row-wise) (Elston and Nicholson 1942).

269	321	585	871	1475	2821	3928	5943	4950	2577
523	98	184	279	409	2285	2685	3409	1824	409
151	45	68	213	546	1033	2129	2536	957	361
377	225	360	731	1638	2725	2871	2119	684	299
236	245	552	1623	3311	6721	4254	687	255	473
358	784	1594	1676	2251	1426	756	299	201	229
469	736	2042	2811	4431	2511	389	73	39	49
59	188	377	1292	4031	3495	587	105	153	387
758	1307	3465	6991	6313	3794	1836	345	382	808
1388	2713	3800	3091	2985	3790	674	81	80	108
229	399	1132	2432	3574	2935	1537	529	485	662
1000	1590	2657	3396						

Appendix B
Matrix approach to the general linear model

B.1 Definition of the general linear model

Many different statistical models can be used to investigate the relation-ship between the distribution of a random *response*, Y, and one or more non-random *explanatory variables* $x_1, x_2, \ldots, x_p$. In the *general linear model*, the mean value of Y is specified as a linear combination of the explanatory variables, the coefficients of this linear combination repre-senting unknown *parameters* to be estimated, whilst the random variation in Y arises by an additive *deviation* from the mean. Algebraically, we can express this as

$$Y = \sum_{j=1}^{p} \theta_j x_j + Z \qquad (\text{B.1})$$

where the *deviation*, Z, is a random variable with zero mean and variance σ^2. Note in particular that the distribution of Z does not depend on the θ_j.

If we now imagine n copies of (B.1), but with varying values of the explanatory variables, we obtain a set of equations of the form

$$Y_i = \sum_{j=1}^{p} \theta_j x_{ij} + Z_i \quad : \quad i = 1, \ldots, n, \qquad (\text{B.2})$$

where now x_{ij} denotes the value of the jth explanatory variable in the ith copy of (B.1). In what follows we shall assume that there are no linear dependencies amongst the explanatory variables. A linear dependency exists if for some fixed coefficients a_j, the linear combination $\sum_{j=1}^{p} a_j x_{ij}$ is zero for all $i = 1, \ldots, n$. Should this be the case, one of the explanatory variables is redundant in the sense that there is an equivalent formulation to (B.2) which uses only $p - 1$ explanatory variables.

A more compact way to express (B.2) is in matrix notation,

$$\mathbf{Y} = X\mathbf{\theta} + \mathbf{Z}, \tag{B.3}$$

where now $\mathbf{Y}$ and $\mathbf{Z}$ are n-element vectors, $\mathbf{\theta}$ is a p-element vector and X is an n by p matrix with (i, j)th element x_{ij}. Equation (B.3) defines the *general linear model*. By suitable choice of explanatory variables, it can encompass a wide variety of standard and non-standard statistical models, including those underlying the analysis of variance for a completely randomized design, analysis of covariance, multiple regression and polynomial regression.

Example (B.1). Parallel quadratic regressions
Suppose that we measure a response, Y, at each of four levels of an explanatory variable, x, and in each of two experimental groups, giving eight observations in all. Suppose also that the mean response in each group is a quadratic function of x, and that the mean response curves in the two groups are parallel. In algebraic terms, the mean response in group k is

$$\alpha_k + \beta x + \gamma x^2$$

for each of $k = 1$ and 2.

The above scenario defines the following general linear model. The number of observations of the response is $n = 8$. The number of explanatory variables is $p = 4$, one corresponding to each element of the vector of parameters, $\mathbf{\theta} = (\alpha_1, \alpha_2, \beta, \gamma)'$. Writing x_1, x_2, x_3, and x_4 for the four levels of the explanatory variable, the matrix X takes the form

$$
X =
\begin{matrix}
1 & 0 & x_1 & x_1^2 \\
1 & 0 & x_2 & x_2^2 \\
1 & 0 & x_3 & x_3^2 \\
1 & 0 & x_4 & x_4^2 \\
0 & 1 & x_1 & x_1^2 \\
0 & 1 & x_2 & x_2^2 \\
0 & 1 & x_3 & x_3^2 \\
0 & 1 & x_4 & x_4^2
\end{matrix}
$$

Note that we use two indicator variables to identify the two experimental groups and construct two further explanatory variables, x and x^2, which are functionally but not linearly dependent. Note also that other parameterizations of the model are possible. For example, if we define

$$X = \begin{matrix} 1 & 0 & x_1 & x_1^2 \\ 1 & 0 & x_2 & x_2^2 \\ 1 & 0 & x_3 & x_3^2 \\ 1 & 0 & x_4 & x_4^2 \\ 1 & 1 & x_1 & x_1^2 \\ 1 & 1 & x_2 & x_2^2 \\ 1 & 1 & x_3 & x_3^2 \\ 1 & 1 & x_4 & x_4^2 \end{matrix}$$

then the parameters α_1, β and γ define the mean response in group 1 as before, but now α_2 represents the *difference* between the mean response curves in the two groups, a quantity which will often be of interest in its own right.

B.2 The multivariate Normal distribution

An n-element random vector $\mathbf{Y}$ has a *multivariate Normal distribution* if any linear combination $\sum_{i=1}^{n} a_i Y_i$ is Normally distributed. In particular, this implies that each Y_i is Normally distributed, although the converse is false: Normality of each Y_i does not guarantee multivariate Normality. The probability density function of the multivariate Normal distribution is

$$f(\mathbf{y}) = (2\pi)^{-n/2} |\Sigma|^{-1/2} \exp\{-\tfrac{1}{2}(\mathbf{y} - \boldsymbol{\mu})' \Sigma^{-1} (\mathbf{y} - \boldsymbol{\mu})\}. \tag{B.4}$$

In (B.4), $\boldsymbol{\mu}$ denotes the *mean of* $\mathbf{Y}$, an n-element vector with elements $\mu_i = E(Y_i)$, and Σ is the *variance* of Y, an n by n symmetric, positive–definite matrix with elements $\Sigma_{ij} = \text{cov}(Y_i, Y_j)$. A convenient shorthand notation for this is

$$\mathbf{Y} \sim MVN(\boldsymbol{\mu}, \Sigma). \tag{B.5}$$

Multivariate Normality is preserved under linear transformation of $\mathbf{Y}$. Specifically, if A is an r by n full-rank matrix with $r < n$, $\mathbf{b}$ is an r-element vector, $\mathbf{Y} \sim MVN(\mu, \Sigma)$ and $\mathbf{U} = A\mathbf{Y} + \mathbf{b}$, then

$$\mathbf{U} \sim MVN(A\boldsymbol{\mu} + \mathbf{b}, A\Sigma A'). \tag{B.6}$$

Finally, in this section, we state a useful result about quadratic forms in multivariate Normal random variables. If $\mathbf{Y} \sim MVN(\boldsymbol{\mu}, \Sigma)$, then

$$(\mathbf{Y} - \boldsymbol{\mu})' \Sigma^{-1} (\mathbf{Y} - \boldsymbol{\mu}) \sim \chi_n^2. \tag{B.7}$$

B.3 Least squares estimation

In applications of the general linear model, we obtain a vector $\mathbf{y}$ of observed responses from which we want to make inferences about $\boldsymbol{\theta}$ within the framework of the model (B.3). If we knew the value of $\boldsymbol{\theta}$, we could compute the vector of observed deviations, $\mathbf{z} = \mathbf{y} - X\boldsymbol{\theta}$. When $\boldsymbol{\theta}$ is unknown, an untuitively sensible way to proceed is to consider $\mathbf{z}$ as a function of $\boldsymbol{\theta}$ and to estimate $\boldsymbol{\theta}$ as the value, $\hat{\boldsymbol{\theta}}$ say, which minimizes $\mathbf{z}'\mathbf{z} = \sum_{i=1}^{n} z_i^2$, the sum of squared deviations. For obvious reasons, this method of estimation is known as *least squares*.

It turns out that $\hat{\boldsymbol{\theta}}$ can be represented explicitly as

$$\hat{\boldsymbol{\theta}} = (X'X)^{-1}X'\mathbf{y}. \tag{B.8}$$

Given $\hat{\boldsymbol{\theta}}$, the vector of *least squares residuals* is defined as

$$\hat{\mathbf{z}} = \mathbf{y} - X\hat{\boldsymbol{\theta}} = \{I - X(X'X)^{-1}X'\}\mathbf{y}. \tag{B.9}$$

These residuals contain information about the variance structure of the response vector, and about the extent to which the model provides a good fit to the data.

We call $\mathbf{y}'\mathbf{y}$ the *total sum of squares* of the original data, $\hat{\mathbf{z}}'\hat{\mathbf{z}}$ the *residual sum of squares* and the difference between these two quantities, $\mathbf{y}'\mathbf{y} - \hat{\mathbf{z}}'\hat{\mathbf{z}}$, the *regression sum of squares* for the fitted model. It follows from (B.9) that we can write the regression sum of squares as

$$\mathbf{y}'X(X'X)^{-1}X'\mathbf{y}. \tag{B.10}$$

The regression sum of squares measures the amount of variation in the original data $\mathbf{y}$ which is explained by fitting the linear model. The residual sum of squares measures the amount of unexplained variation. This unexplained variation may be purely random, in which case the model fits the data well, or it may include a systematic component due to lack of fit. For example, a relevant explanatory variable may have been omitted, or the true relationship between the mean response and the explanatory variables may be non-linear.

B.4 Properties of least squares estimates under a Normality assumption

In order to make formal inferences about $\boldsymbol{\theta}$ in the general linear model (B.3), we need to adopt a distributional model for the deviations $\mathbf{Z}$. We shall assume here that the Z_i are independent and Normally distributed, with common variance σ^2. This constitutes a special case of the multivariate Normal distribution, namely

$$\mathbf{Z} \sim MVN(0, \sigma^2 I),$$

where I denotes the n by n identity matrix. It follows from a simple application of (B.6) that

$$\mathbf{Y} \sim MVN(X\mathbf{\theta}, \sigma^2 I). \tag{B.11}$$

Now, (B.8) defines the least squares estimate $\hat{\mathbf{\theta}}$ as a linear transformation of the observed response vector $\mathbf{y}$. The corresponding linear transformation of the random vector $\mathbf{Y}$ is itself a multivariate random variable,

$$\hat{\mathbf{\theta}} = (X'X)^{-1}X'\mathbf{Y},$$

and a second application of (B.6) in conjunction with (B.11) shows that

$$\hat{\mathbf{\theta}} \sim MVN(\mathbf{\theta}, \sigma^2(X'X)^{-1}). \tag{B.12}$$

This last result shows in particular that $\hat{\mathbf{\theta}}$ is an unbiased estimator for $\mathbf{\theta}$ but that, as we might expect, its variance depends on the unknown σ^2.

In fact, $\hat{\mathbf{\theta}}$ is the maximum likelihood estimator for $\mathbf{\theta}$ under the Normality assumption. Furthermore, the maximum likelihood estimate of σ^2 is the average squared residual,

$$\hat{\sigma}^2 = n^{-1}\hat{\mathbf{z}}'\hat{\mathbf{z}}.$$

Approximate inferences about $\mathbf{\theta}$ can be made using (B.12) but with $\hat{\sigma}^2$ in place of σ^2. For example, approximate standard errors for the $\hat{\theta}_j$ can be computed as the square roots of the diagonal elements of $\hat{\sigma}^2(X'X)^{-1}$, whilst the off-diagonal elements of this matrix give information about correlations amongst the $\hat{\theta}_j$.

Exact inferences are also possible, using the fact that $\hat{\sigma}^2$ is independent of $\hat{\theta}$ and has a distribution proportional to χ^2_{n-p}. For example, exact inferences about an individual parameter θ_j can be based on the statistic

$$(\hat{\theta}_j - \theta_j)/\sqrt{\{n\hat{\sigma}/(n - p)\}},$$

whose sampling distribution is Student's t on $n - p$ degrees of freedom. The slightly complicated form of the denominator of this statistic arises because $\hat{\sigma}^2$ is a biased estimator of σ^2. An unbiased estimator is $\tilde{\sigma}^2 = n\hat{\sigma}^2/(n - p)$.

B.5 Generalized least squares

In some applications, particularly those involving time-series data, the assumption of uncorrelated X_i is unreasonable. A more general model is to assume that

$$\mathbf{Z} \sim MVN(\mathbf{0}, \sigma^2 V). \tag{B.13}$$

In (B.13) we write the variance of $\mathbf{Z}$ as $\sigma^2 V$ to emphasize that σ^2 is typically unknown. The elements of V may also involve unknown parameters, but we shall assume not for the moment.

It follows from (B.13) and (B.6) that

$$\mathbf{Y} \sim MVN(X\boldsymbol{\theta}, \sigma^2 V). \tag{B.14}$$

Under this model, the maximum likelihood estimator for $\boldsymbol{\theta}$ is

$$\hat{\boldsymbol{\theta}} = (X'V^{-1}X)^{-1}X'V^{-1}\mathbf{y}. \tag{B.15}$$

Again using (B.6), we deduce that

$$\hat{\boldsymbol{\theta}} \sim MVN(\boldsymbol{\theta}, \sigma^2(X'V^{-1}X)^{-1}). \tag{B.16}$$

Note in particular that when $V = I$, (B.15) and (B.16) reduce to (B.8) and (B.12), respectively. The maximum likelihood estimator for σ^2 is

$$\hat{\sigma}^2 = n^{-1}\hat{\mathbf{z}}V^{-1}\hat{\mathbf{z}}, \tag{B.17}$$

where

$$\hat{\mathbf{z}} = \mathbf{y} - X\hat{\boldsymbol{\theta}}$$

as before. Approximate inferences about $\boldsymbol{\theta}$ follow from (B.16) with $\hat{\sigma}^2$ in place of σ^2. For exact inferences, we again use the fact that $\hat{\sigma}^2$ is independent of $\hat{\boldsymbol{\theta}}$ and has a distribution proportional to χ^2_{n-p}.

If the matrix V involves unknown parameters, $\boldsymbol{\phi}$ say, then (B.15) and (B.17) are both functions of $\boldsymbol{\phi}$. If we substitute these expressions into the formula for the multivariate Normal density (B.4), we obtain a likelihood function for $\boldsymbol{\phi}$ from which we can derive the maximum likelihood estimator $\hat{\boldsymbol{\phi}}$. Typically, we have to resort to numerical methods to maximize this reduced likelihood function. Having done so, we can substitute $\hat{\boldsymbol{\phi}}$ into (B.15) and (B.17) to obtain the maximum likelihood estimators $\hat{\boldsymbol{\theta}}$ and $\hat{\sigma}^2$.

In this extended framework , exact inferences about $\boldsymbol{\theta}$ are typically no longer available. However, approximate inferences can still be based on (B.16), but with the maximum likelihood estimators $\hat{\sigma}^2$ and $\hat{\boldsymbol{\phi}}$ used in place of the unknown σ^2 and $\boldsymbol{\phi}$.

B.6   Further reading

There are many books giving comprehensive treatments of the general linear model. Two good examples are Draper and Smith (1981) and Seber (1977), the latter being somewhat the more demanding mathematically.

Chatfield and Collins (1980) and Mardia *et al.* (1979) are but two of many books which include detailed accounts of the multivariate Normal distribution, the latter again being the more demanding.

Appendix C
Complex numbers

C.1 Definitions

Complex numbers arise in the context of quadratic equations which have no real solution. For example, the quadratic equation

$$z^2 - 5z + 4 = 0$$

has two real solutions, $z = 4$ and $z = 1$, whereas the equation

$$z^2 + 3z + 4 = 0$$

has none. The general quadratic equation

$$az^2 + bz + c = 0$$

has solutions

$$z = \{-b \pm \sqrt{(b^2 - 4ac)}\}/(2a)$$

provided $b^2 - 4ac \geqslant 0$. A possible way out of this dilemma is to define an *imaginary number* i with the convenient property that $i^2 = -1$. It turns out that if we do this, all quadratic equations have solutions.

We define a *complex number* z to be any number of the form $z = x + iy$, where x and y are real numbers and $i^2 = -1$. We call x and y the *real* and *imaginary parts* of z. To add two complex numbers together, we add real and imaginary parts separately. Thus, if $z_1 = x_1 + iy_1$, and $z_2 = x_2 + iy_2$, then $z_1 + z_2 = (x_1 + x_2) + i(y_1 + y_2)$. To multiply two complex numbers, we operate exactly as we do in multiplying any two algebraic expressions. Thus, with z_1 and z_2 as before, we obtain

$$z_1 z_2 = (x_1 + iy_1)(x_2 + iy_2)$$
$$= x_1 x_2 + i(y_1 x_2 + x_1 y_2) + i^2 y_1 y_2$$
$$= (x_1 x_2 - y_1 y_2) + i(x_1 y_2 + y_1 x_2),$$

because $i^2 = -1$. Note that if $x_2 = x_1$ and $y_2 = -y_1$, then $z_1 z_2$ is the positive real number $x_1^2 + y_1^2$. We call $\bar{z} = x - iy$ the *complex conjugate* of $z = x + iy$, and vice versa. Also, we define the *modulus* of z to be $|z| = \sqrt{(x^2 + y^2)}$. Finally, we note that if $z_1 = x_1 + iy_1$, and $z_2 = \bar{z}_1/|z_1|^2$, ·

252 APPENDIX C

then

$$z_1 z_2 = (x_1 + iy_1)(x_1 - iy_1)/(x_1^2 + y_1^2) = 1.$$

We call $z^{-1} = \bar{z}/|z|^2$ the *inverse* of z, and define complex division as complex multiplication by the inverse, i.e. $z_1/z_2 = z_1 z_2^{-1}$.

C.2 Polar representation of complex numbers

Clearly, there is an exact correspondence between complex numbers z and *pairs* of real numbers (x, y). If we think of any such (x, y) as a point in the Euclidean plane, we can equally represent it by its distance from the origin, $r = \sqrt{(x^2 + y^2)}$, and the angle θ between the x-axis and a line joining the origin to the point (x, y). This gives

$$x = r \cos \theta$$

and

$$y = r \sin \theta,$$

which in turn gives us the alternative representation of z as

$$z = r(\cos \theta + i \sin \theta). \tag{C.1}$$

Note that in this representation $r = |z|$, because all complex numbers of the form $\cos \theta + i \sin \theta$ have modulus $\sqrt{(\cos^2 \theta + \sin^2 \theta)} = 1$.

C.3 The complex exponential function

The polar representation (C.1) make complex multiplication extremely easy. Let $z_1 = r_1(\cos \theta_1 + i \sin \theta_1)$ and $z_2 = r_2(\cos \theta_2 + i \sin \theta_2)$. Then,

$$\begin{aligned} z_1 z_2 &= r_1 r_2 (\cos \theta_1 + \sin \theta_1)(\cos \theta_2 + i \sin \theta_2) \\ &= r_1 r_2 \{(\cos \theta_1 \cos \theta_2 - \sin \theta_1 \sin \theta_2) + i(\sin \theta_1 \cos \theta_2 + \cos \theta_1 \sin \theta_2)\} \\ &= r_1 r_2 (\cos(\theta_1 + \theta_2) + i \sin(\theta_1 + \theta_2)). \end{aligned}$$

In other words, to multiply any two complex numbers we simply multiply moduli and add angles. In real analysis, a similar rule applies to the multiplication of two exponentials, $a e^{x_1} b e^{x_2} = ab e^{x_1 + x_2}$. This prompts us to *define* the exponential function

$$e^{i\theta} = \cos \theta + i \sin \theta, \tag{C.2}$$

allowing us to write the polar representation of a complex number in the compact form $r e^{i\theta}$. Note in particular that $r e^{i\theta}$ is a real number whenever θ is an integer multiple of π. Also, (C.2) shows that $\cos \theta = \frac{1}{2}(e^{i\theta} + e^{-i\theta})$.

Writing complex numbers in this way makes standard algebraic manipulations relatively straightforward. For example, suppose that we want to sum the first n terms of a geometric series,

$$S_n = \sum_{k=0}^{n-1} z^k.$$

Using a standard result from the elementary theory of real series, we can write

$$S_n = (1 - z^n)/(1 - z), \tag{C.3}$$

provided $z \neq 1$. Writing $z = re^{i\theta}$, (C.3) becomes

$$S_n = (1 - r^n e^{in\theta})/(1 - re^{i\theta}). \tag{C.4}$$

In particular, if $r = 1$ and $n\theta$ is an integer multiple of 2π, (C.4) gives $S_n = 0$.

C.4 Further reading

Ledermann (1960) is a brief introduction to the theory of complex numbers. Hawkins and Hawkins (1969) is a more detailed, but still elementary, treatment. For the reader who wishes to explore the full sophistication of complex analysis, which deals with general properties of functions of a complex variable, available texts include Copson (1935), Jameson (1970), and Priestley (1985).

Author index

Subject index